The Universe
in a Single Atom

HIS HOLINESS THE

Dalai Lama

THE UNIVERSE IN A SINGLE ATOM

THE CONVERGENCE OF SCIENCE AND SPIRITUALITY

BROADWAY BOOKS

New York

BROADWAY

PUBLISHED BY BROADWAY BOOKS

A hardcover edition of this book was originally published in 2005
by Broadway Books.
Published in the United States by Broadway Books, an imprint of
The Doubleday Broadway Publishing Group, a division of Random House,
Inc., New York.
www.broadwaybooks.com

BROADWAY BOOKS and its logo, a letter B bisected on the diagonal, are trademarks
of Random House, Inc.

Library of Congress Cataloging-in-Publication Data
Bstan-'dzin-rgya-mtsho, Dalai Lama XIV, 1935–
    The universe in a single atom : the convergence of science and
    spirituality / the Dalai Lama.
        p.   cm.
    1. Buddhism and science.   I. Title.
    BQ7935.B774U56 2005
    294.3'365—dc22
                                                            2005049556

ISBN-13: 978-0-7679-2081-0
ISBN-10: 0-7679-2081-3

PRINTED IN THE UNITED STATES OF AMERICA

20 19 18 17 16 15 14 13 12 11

In each atom of the realms of the universe,
There exist vast oceans of world systems.

The Great Flower Ornament,

an ancient Buddhist scripture

Contents

* * * * *

PROLOGUE

◆　◆　◆　◆　◆

I was never myself trained in science. My knowledge comes mainly from reading news coverage of important scientific stories in magazines like *Newsweek*, or hearing reports on the BBC World Service and later reading textbooks on astronomy. Over the last thirty years I have held many personal meetings and discussions with scientists. In these encounters, I have always attempted to grasp the underlying models and methods of scientific thought as well as the implications of particular theories or new discoveries. But I have nonetheless thought deeply about science—not just its implications for the understanding of what reality is but the still more important question of how it may influence ethics and human values. The specific areas of science that I have explored most over the years are subatomic physics, cosmology, and biology, including neuroscience and psychology. Given that my own intellectual training is in Buddhist thought, naturally I have often wondered about the interface of key

Buddhist concepts and major scientific ideas. This book is a result of that long period of thinking and of the intellectual journey of a Buddhist monk from Tibet into the world of bubble chambers, particle accelerators, and fMRI (functional magnetic resonance imaging).

Many years after I went into exile in India, I came across an open letter from the 1940s addressed to the Buddhist thinkers of Tibet. It was written by Gendün Chöphel, a Tibetan scholar who not only had mastered Sanskrit but also, uniquely among Tibetan thinkers of his time, had a good command of English. He traveled extensively in British India, Afghanistan, Nepal, and Sri Lanka in the 1930s. This letter, composed toward the end of his twelve-year trip, was amazing to me. It articulates many of the areas in which there could be a fruitful dialogue between Buddhism and modern science. I discovered that Gendün Chöphel's observations often coincide remarkably with my own. It is a pity that this letter did not attract the attention it deserved, partly because it was never properly published in Tibet before I came into exile in 1959. But I find it heartwarming that my journey into the scientific world has a precedent within my own Tibetan tradition. All the more so since Gendün Chöphel came from my native province of Amdo. Encountering this letter so many years after it was written was an impressive moment.

I remember a disturbing conversation I had had only a few years earlier with an American lady who was married to a Tibetan. Having heard of my interest in science and my active engagement in dialogue with scientists, she warned me of the danger science poses to the survival of Buddhism. She told me that history attests to the fact that science is the "killer" of religion and advised me that it was not wise for the Dalai Lama to pursue friendships with those who represent this profession. By taking this personal journey into science, I suppose I have stuck my neck out. My confidence in venturing into science lies in my basic belief that as in science so in

Buddhism, understanding the nature of reality is pursued by means of critical investigation: if scientific analysis were conclusively to demonstrate certain claims in Buddhism to be false, then we must accept the findings of science and abandon those claims.

Because I am an internationalist at heart, one of the qualities that has moved me most about scientists is their amazing willingness to share knowledge with each other without regard for national boundaries. Even during the Cold War, when the political world was polarized to a dangerous degree, I found scientists from the Eastern and Western blocs willing to communicate in ways the politicians could not even imagine. I felt an implicit recognition in this spirit of the oneness of humanity and a liberating absence of proprietorship in matters of knowledge.

The motivation for my interest in science is more than merely personal. Even before I came into exile, it was clear to me and others in the country that one of the underlying causes for Tibet's political tragedy was its failure to open itself to modernization. As soon as we arrived in India, we set up Tibetan schools for refugee children with a modern curriculum, which included scientific education for the first time. By then I had come to recognize that the essence of modernization lies in the introduction of modern education, and at the heart of modern education there must be a command of science and technology. My personal commitment to this educational project has led me to encourage even the monastic colleges, whose primary role is to teach classical Buddhist thought, to introduce science into their curriculum.

As my comprehension of science has grown, it has gradually become evident to me that, insofar as understanding the physical world is concerned, there are many areas of traditional Buddhist thought where our explanations and theories are rudimentary when compared with those of modern science. But at the same time, even in the most highly developed scientific countries, it is clear that human beings continue to experience suffering, especially at the emotional and psychological level. The great benefit of science is that it can contribute tremendously to the alleviation of suffering at the physical level, but it is only through the cultivation of the qualities of the human heart and the transformation of our attitudes that we can begin to address and overcome our mental suffering. In other words, the enhancement of fundamental human values is indispensable to our basic quest for happiness. Therefore, from the perspective of human well-being, science and spirituality are not unrelated. We need both, since the alleviation of suffering must take place at both the physical and the psychological levels.

This book is not an attempt to unite science and spirituality (Buddhism being the example I know best) but an effort to examine two important human disciplines for the purpose of developing a more holistic and integrated way of understanding the world around us, one that explores deeply the seen and the unseen, through the discovery of evidence bolstered by reason. I am not attempting a scholarly treatment of the potential points of convergence and difference between Buddhism and science—I leave that to professional academics. Rather, I believe that spirituality and science are different but complementary investigative approaches with the same greater goal, of seeking the truth. In this, there is much each may learn from the other, and together they may contribute to expanding the horizon of human knowledge and wisdom. Moreover, through a dialogue between the two disciplines, I hope both science and spirituality may develop to be of better service to the needs and well-being of humanity. In addition, by telling the story of my own journey, I wish to emphasize to the millions of my fellow Buddhists worldwide the need to take science seriously and to accept its fundamental discoveries within their worldview.

This dialogue between science and spirituality has a long history—especially with respect to Christianity. In the case of my own tradition, Tibetan Buddhism, for various historical, social, and political reasons, the full encounter with a scientific worldview is still a novel process. The implications of what science has to offer are still not wholly clear. Regardless of different personal views about science, no credible understanding of the natural world or our human existence—what I am going to call in this book a worldview—can ignore the basic insights of theories as key as evolution, relativity, and quantum mechanics. It may be that science will learn from an engagement with spirituality, especially in its interface with wider human issues, from ethics to society, but certainly some specific aspects of Buddhist thought—such as its old cosmological theories and its rudimentary physics—will have to be modified in the light of new scientific insights. I hope this book will be a contribution to the critical project of enlivening the dialogue between science and spirituality.

Because my aim is to explore issues of the deepest significance for our contemporary world, I have wished to communicate with the widest possible audience. This is not easy, given the sometimes complex reasoning and argumentation in both science and Buddhist philosophy. In my eagerness to make the discussion accessible, I may on occasion have oversimplified issues. I am grateful to my two editors, my longtime translator Thupten Jinpa and his colleague Jas' Elsner, for their assistance in helping me to articulate my ideas as lucidly as possible in English. I also wish to thank the numerous individuals who have helped them and commented on the various stages of the manuscript. Above all, I am grateful to all the scientists who have met with me, been so generous with their time, and shown such extraordinary patience in explaining complex ideas to a sometimes slow student. I regard them all as my teachers.

I

.

REFLECTION

I have spent many years reflecting on the remarkable advances of science. Within the short space of my own lifetime, the impact of science and technology on humanity has been tremendous. Although my own interest in science began with curiosity about a world, foreign to me at that time, governed by technology, it was not very long before the colossal significance of science for humanity as a whole dawned on me—especially after I came into exile in 1959. There is almost no area of human life today that is not touched by the effects of science and technology. Yet are we clear about the place of science in the totality of human life—what exactly it should do and by what it should be governed? This last point is critical because unless the direction of science is guided by a consciously ethical motivation, especially compassion, its effects may fail to bring benefit. They may indeed cause great harm.

Seeing the tremendous importance of science and recognizing

its inevitable dominance in the modern world fundamentally changed my attitude to it from curiosity to a kind of urgent engagement. In Buddhism the highest spiritual ideal is to cultivate compassion for all sentient beings and to work for their welfare to the greatest possible extent. From my earliest childhood I have been conditioned to cherish this ideal and attempt to fulfill it in my every action. So I wanted to understand science because it gave me a new area to explore in my personal quest to understand the nature of reality. I also wanted to learn about it because I recognized in it a compelling way to communicate insights gleaned from my own spiritual tradition. So, for me, the need to engage with this powerful force in our world has become a kind of spiritual injunction as well. The central question—central for the survival and well-being of our world—is how we can make the wonderful developments of science into something that offers altruistic and compassionate service for the needs of humanity and the other sentient beings with whom we share this earth.

Do ethics have a place in science? I believe they do. First of all, like any instrument, science can be put to good use or bad. It is the state of mind of the person wielding the instrument that determines to what end it will be put. Second, scientific discoveries affect the way we understand the world and our place in it. This has consequences for our behavior. For example, the mechanistic understanding of the world led to the Industrial Revolution, in which the exploitation of nature became the standard practice. There is, however, a general assumption that ethics are relevant to only the application of science, not the actual pursuit of science. In this model the scientist as an individual and the community of scientists in general occupy a morally neutral position, with no responsibility for the fruits of what they have discovered. But many important scientific discoveries, and particularly the technological innovations they lead to, create new conditions and open up new possibilities which give rise to new ethical and spiritual challenges. We cannot simply absolve the scientific enterprise and individual scientists from responsibility for contributing to the emergence of a new reality.

Perhaps the most important point is to ensure that science never becomes divorced from the basic human feeling of empathy with our fellow beings. Just as one's fingers can function only in relation to the palm, so scientists must remain aware of their connection to society at large. Science is vitally important, but it is only one finger of the hand of humanity, and its greatest potential can be actualized only so long as we are careful to remember this. Otherwise, we risk losing our sense of priorities. Humanity may end up serving the interests of scientific progress rather than the other way around. Science and technology are powerful tools, but we must decide how best to use them. What matters above all is the motivation that governs the use of science and technology, in which ideally heart and mind are united.

For me, science is first and foremost an empirical discipline that provides humanity with a powerful access to understanding the nature of the physical and living world. It is essentially a mode of inquiry that gives us fantastically detailed knowledge of the empirical world and the underlying laws of nature, which we infer from the empirical data. Science proceeds by means of a very specific method that involves measurement, quantification, and intersubjective verification through repeatable experiments. This, at least, is the nature of scientific method as it exists within the current paradigm. Within this model, many aspects of human existence, including values, creativity, and spirituality, as well as deeper metaphysical questions, lie outside the scope of scientific inquiry.

Though there are areas of life and knowledge outside the domain of science, I have noticed that many people hold an assumption that the scientific view of the world should be the basis for all knowledge and all that is knowable. This is scientific materialism. Although I am not aware of a school of thought that explicitly propounds this notion, it seems to be a common unexamined presupposition. This view upholds a belief in an objective world, independent of the contingency of its observers. It assumes that the data being analyzed within an experiment are independent of the preconceptions, perceptions, and experience of the scientist analyzing them.

Underlying this view is the assumption that, in the final analysis, matter, as it can be described by physics and as it is governed by the laws of physics, is all there is. Accordingly, this view would uphold that psychology can be reduced to biology, biology to chemistry, and chemistry to physics. My concern here is not so much to argue against this reductionist position (although I myself do not share it) but to draw attention to a vitally important point: that these ideas do not constitute scientific knowledge; rather they represent a philosophical, in fact a metaphysical, position. The view that all aspects of reality can be reduced to matter and its various particles is, to my mind, as much a metaphysical position as the view that an organizing intelligence created and controls reality.

One of the principal problems with a radical scientific materialism is the narrowness of vision that results and the potential for nihilism that might ensue. Nihilism, materialism, and reductionism are above all problems from a philosophical and especially a human perspective, since they can potentially impoverish the way we see ourselves. For example, whether we see ourselves as random biological creatures or as special beings endowed with the dimension of consciousness and moral capacity will make an impact on how we feel about ourselves and treat others. In this view many dimensions of the full reality of what it is to be human—art, ethics, spirituality, goodness, beauty, and above all, consciousness—either are reduced to the chemical reactions of firing neurons or are seen as a matter of purely imaginary constructs. The danger then is that human beings may be reduced to nothing more than biological machines, the products of pure chance in the random combination of genes, with no purpose other than the biological imperative of reproduction.

It is difficult to see how questions such as the meaning of life or good and evil can be accommodated within such a worldview. The problem is not with the empirical data of science but with the contention that these data alone constitute the legitimate ground for developing a comprehensive worldview or an adequate means for responding to the world's problems. There is more to human existence and to reality itself than current science can ever give us access to.

By the same token, spirituality must be tempered by the insights and discoveries of science. If as spiritual practitioners we ignore the discoveries of science, our practice is also impoverished, as this mind-set can lead to fundamentalism. This is one of the reasons I encourage my Buddhist colleagues to undertake the study of science, so that its insights can be integrated into the Buddhist worldview.

2.

ENCOUNTER
WITH SCIENCE

I was born into a family of simple farmers who used cattle to plow their field and, when the barley was harvested, used cattle to trample the grain out of the husk. Perhaps the only objects that could be described as technological in the world of my early childhood were the rifles that local warrior nomads had probably acquired from British India, Russia, or China. At the age of six I was enthroned as the Fourteenth Dalai Lama in the Tibetan capital, Lhasa, and embarked upon an education in all aspects of Buddhism. I had personal tutors who gave me daily classes in reading, writing, basic Buddhist philosophy, and the memorization of scriptures and rituals. I was also given several *tsenshap*, which literally means "philosophical assistants." Their primary job was to engage me in debate on issues of Buddhist thought. In addition, I would participate in long hours of prayers and meditative contemplation. I spent periods in retreat with my tutors and sat regularly for two

hours at a time four times a day in a meditation session. This is a
fairly typical training for a high lama in the Tibetan tradition. But
I was not educated in math, geology, chemistry, biology, or physics.
I did not even know they existed.

The Potala Palace was my official winter residence. It is a huge
edifice, occupying the entire side of a mountain, and is supposed
to have a thousand rooms—I never counted them myself. In my
spare moments as a boy, I amused myself by exploring some of its
chambers. It was like being on a perpetual treasure hunt. There
were all kinds of things, mainly the belongings of former Dalai
Lamas and especially of my immediate predecessor, preserved
there. Among the most striking of the palace's contents were the
reliquary stupas containing the remains of the previous Dalai
Lamas, reaching back to the Fifth, who lived in the seventeenth
century and enlarged the Potala to its present form. Amid the as-
sorted oddities I found lying about were some mechanical objects
which belonged to the Thirteenth Dalai Lama. Most notable were
a collapsible telescope made from brass, which could be attached
to a tripod, and a hand-wound mechanical timepiece with a rotat-
ing globe on a stand that gave the time in different time zones.
There was also a large stash of illustrated books in English telling
the story of the First World War.

Some of these were gifts to the Thirteenth Dalai Lama from his
friend Sir Charles Bell. Bell was the Tibetan-speaking British political officer in Sikkim. He had been the Thirteenth Dalai Lama's
host during his brief sojourn in British India when he fled in 1910
at the threat of invasion by the armies of the last imperial government of China. It is curious that exile in India and the discovery of
scientific culture are things bequeathed to me by my most immediate predecessor. For the Thirteenth Dalai Lama, as I later found
out, this stay in British India was an eye-opening experience, which

led to a recognition of the need for major social and political reforms in Tibet. On his return to Lhasa, he introduced the telegraph, set up a postal service, built a small generating plant to power Tibet's first electric lights, and established a mint for the national coinage and the printing of paper currency. He also came to appreciate the importance of a modern, secular education and sent a select group of Tibetan children to study at Rugby School in England. The Thirteenth Dalai Lama left a remarkable deathbed testament, which predicted much of the political tragedy to come and which the government that succeeded him failed to understand fully or to heed.

Among the other items of mechanical interest acquired by the Thirteenth Dalai Lama were a pocket watch, two film projectors, and three motorcars—two Baby Austins from 1927 and a 1931 American Dodge. As there were no drivable roads across the Himalayas or in Tibet itself, these cars had to be disassembled in India and carried across the mountains by porters, mules, and donkeys before being put back together again for the Thirteenth Dalai Lama. For a long time these were the only three automobiles in all Tibet—and pretty useless they were, since there were no roads outside Lhasa on which one could drive them. These various items, the telltale signs of a technological culture, exercised great fascination on a naturally curious and somewhat restless boy. There was a time, I remember very clearly, when I would rather fiddle with these objects than study philosophy or memorize a text. Today I can see that these things were in themselves no more than toys, but they hinted at a whole universe of experience and knowledge to which I had no access and whose existence was endlessly tantalizing. In a way, this book is about the path to discovering that world and the wonderful things it has to offer.

I did not find the telescope a problem. Somehow it was quite obvious to me what it was for, and I was soon using it to observe the bustling life of Lhasa town, especially the marketplaces. I envied the sense of abandon with which children of my age could run about in the streets while I had to study. Later I used the telescope to peer into the night sky above the Potala—which offers, in the high altitude of Tibet, one of the most spectacular views of the stars. I asked my attendants the names of the stars and constellations.

I knew what the pocket watch was for but was much more intrigued by how it worked. I puzzled over this for some time, until curiosity got the better of me and I opened up the case to look inside. Soon I had dismantled the entire item, and the challenge was to put it back together again so that it actually worked. Thus began what was to become a lifelong hobby of dismantling and reassembling mechanical objects. I mastered this process well enough to become the principal repairer for a number of the people I knew who owned watches or clocks in Lhasa. In India later on, I did not have much luck with my cuckoo clock, whose poor cuckoo got attacked by my cat and never recovered. When the automatic battery watch became common, my hobby got much less interesting—if you open one of these, you find hardly any mechanism at all.

Figuring out how to use the Thirteenth Dalai Lama's two hand-cranked film projectors was much more complicated. One of my attendants, an ethnic Chinese monk, worked out how to make one of them function. I asked him to set it up so that I could watch the very few films we had. Later we got hold of a sixteen-millimeter electrically powered projector, but it kept breaking down, partly because the generator which powered it was faulty. Around this time, I guess in 1945, Heinrich Harrer and Peter Aufschnaiter, Austrians who had escaped over the Himalayas from a British prisoner of war camp in northern India, arrived in Lhasa. Harrer became a friend of mine, and I would occasionally turn to him to help fix the projector. We could not get many films, but numerous newsreels of the great events of the Second World War made it across from India, giving the story from an Allied perspective. There were also reels of VE Day, of the coronation of King George VI of England and Laurence Olivier's film of Shakespeare's *Henry V,* as well as some of Charlie Chaplin's silent movies.

My fascination for science began with technology, and indeed I saw no difference between the two. When I met Harrer, who was much better with things mechanical than anyone I knew in Lhasa, I presumed his expertise in science was as profound as his command over the few mechanical objects we had in the Potala. It is funny that years later I discovered he had no professional scientific background—at that time I thought all white men had deep knowledge of science.

Inspired by my success in dismantling watches and repairing the projector, I got more ambitious. My next project was to understand the mechanics of the automobile. The man in charge of driving and looking after the cars was called Lhakpa Tsering; he was a bald fellow whose ill temper was legendary. If he accidentally banged his head while working beneath the car, he would become so angry that he would thrash and bang it again. I made friends with him so that he would allow me to examine the engine while he was repairing it and eventually show me how to drive.

One day I sneaked one of the Austins out for a solo drive but had a small accident and broke the left headlight. I was terrified of what Babu Tashi, another man in charge of the cars, might say. I managed to find a replacement headlight, but it was of clear glass, whereas the original had been frosted. After some thought, I found a solution. I reproduced the light's frosted appearance by covering it with molten sugar. I never knew if Babu Tashi found out. If he did, at least he never punished me.

In one crucial area of modern science, Harrer was most helpful to me; this was world geography. Within my personal library
was a collection of English volumes on the Second World War,
which gave detailed accounts of the participation in the war by a
great many nations, including Japan. My adventures with the movie
projector, fixing clocks, and trying to drive a car gave me an inkling
of what the world of science and technology might be about. On a
more serious level, after I had been invested with the leadership of
Tibet at the age of sixteen, I embarked on state visits to China in
1954 and India in 1956, which left a strong impression. The Chinese army had already invaded my country, and I was involved in a
long and delicate negotiation searching for an accommodation
with the Chinese government.

My first foreign trip, when I was in my late teens, took me to
Beijing, where I met Chairman Mao, Chou En-lai, and other leaders of the regime. This state visit included a series of excursions to
cooperative farms and major utilities such as hydroelectric dams.
Not only was this the first time I was in a modern city with paved
roads and cars but it was also when I first met real scientists.

In 1956 I went to India to take part in the 2,500th anniversary
of the Buddha's death, whose main event took place in Delhi.
Later, the Indian Prime Minister Jawaharlal Nehru became something of a counselor to me and a friend, as well as my host in exile.
Nehru was scientifically minded; he saw India's future in terms of
technological and industrial development and had a profound vision of progress. After the formal celebration of the Buddha's final
passing away, I saw many parts of India—not only the pilgrimage
sites like Bodhgaya, where the Buddha attained full awakening, but
also major cities, industrial complexes, and universities.

It was then that I had my first encounters with spiritual teachers who were seeking the integration of science and spirituality,

such as the members of the Theosophical Society in Madras. Theosophy was an important spiritualist movement in the nineteenth and early twentieth centuries that sought to develop a synthesis of human knowledge, Eastern and Western, religious and scientific. Its founders, including Madame Blavatsky and Annie Besant, were Westerners but spent much time in India.

Even before these official trips, I came to recognize that technology is in fact the fruit, or expression, of a particular way of understanding the world. Science is the basis of these expressions. Science, however, is the specific form of inquiry and the body of knowledge derived from it that give rise to this understanding of the world. So although my initial fascination was with the technological artifacts, it is this—the scientific form of inquiry rather than any particular industry or mechanical toy—that has come to intrigue me most deeply.

As a result of talking to people, especially professional scientists, about science, I noticed certain similarities in the spirit of inquiry between science and Buddhist thought—similarities that I still find striking. The scientific method, as I understand it, proceeds from the observation of certain phenomena in the material world, leads to a theoretical generalization, which predicts the events and results that arise if one treats the phenomena in a particular way, and then tests the prediction with an experiment. The result is accepted as part of the body of wider scientific knowledge if the experiment is correctly conducted and may be repeated. However, if the experiment contradicts the theory, then it is the theory that needs to be adapted—since the empirical observation of phenomena has priority. Effectively, science moves from empirical experience via a conceptual thought process that includes the application of reason and culminates in further empirical experience to verify the understanding offered by reason. I have long been gripped with a fascination for the parallels between this form of empirical investigation and those I had learned in my Buddhist philosophical training and contemplative practice.

Although Buddhism has come to evolve as a religion with a characteristic body of scriptures and rituals, strictly speaking, in Buddhism scriptural authority cannot outweigh an understanding based on reason and experience. In fact the Buddha himself, in a famous statement, undermines the scriptural authority of his own words when he exhorts his followers not to accept the validity of his teachings simply on the basis of reverence to him. Just as a seasoned goldsmith would test the purity of his gold through a meticulous process of examination, the Buddha advises that people should test the truth of what he has said through reasoned examination and personal experiment. Therefore, when it comes to validating the truth of a claim, Buddhism accords greatest authority to experience, with reason second and scripture last. The great masters of the Nalanda school of Indian Buddhism, from which Tibetan Buddhism sprang, continued to apply the spirit of the Buddha's advice in their rigorous and critical examination of the Buddha's own teachings.

In one sense the methods of science and Buddhism are different: scientific investigation proceeds by experiment, using instruments that analyze external phenomena, whereas contemplative investigation proceeds by the development of refined attention, which is then used in the introspective examination of inner experience. But both share a strong empirical basis: if science shows something to exist or to be non-existent (which is not the same as not finding it), then we must acknowledge that as a fact. If a hypothesis is tested and found to be true, we must accept it. Likewise, Buddhism must accept the facts—whether found by science or found by contemplative insights. If, when we investigate something, we find there is reason and proof for it, we must acknowledge that as reality—even if it is in contradiction with a literal scriptural explanation that has held sway for many centuries or with a deeply held opinion or view. So one fundamental attitude shared by Buddhism and science is the commitment to keep searching for reality by empirical means and to be willing to discard accepted or long-held positions if our search finds that the truth is different.

By contrast with religion, one significant characteristic of science is the absence of an appeal to scriptural authority as a source of validating truth claims. All truths in science must be demonstrated either through experiment or through mathematical proof. The idea that something must be so because Newton or Einstein said so is simply not scientific. So an inquiry has to proceed from a state of openness with respect to the question at issue and to what the answer might be, a state of mind which I think of as healthy skepticism. This kind of openness can make individuals receptive to fresh insights and new discoveries; and when it is combined with the natural human quest for understanding, this stance can lead to a profound expanding of our horizons. Of course, this does not mean that all practitioners of science live up to this ideal. Some may indeed be caught in earlier paradigms.

With regard to the Buddhist investigative traditions, we Tibetans owe a tremendous debt to classical India, the birthplace of Buddhist philosophical thinking and spiritual teaching. Tibetans have always referred to India as "the Land of the Noble Ones." This is the country that gave birth to the Buddha, and to a series of great Indian masters whose writings have fundamentally shaped the philosophical thinking and the spiritual tradition of the Tibetan people—the second-century philosopher Nagarjuna, the fourth-century luminaries Asanga and his brother Vasubandhu, the great

ethical teacher Shantideva, and the seventh-century logician Dharmakirti.

Since my flight from Tibet in March 1959, a large number of Tibetan refugees and I have been extremely fortunate to find a second home in India. The president of India in my early years in exile
was Dr. Rajendra Prasad, a deeply spiritual man and respected legal scholar. The vice president, who later became president, was Dr.
Sarvepalli Radhakrishnan, whose professional and personal interests in philosophy were widely known. I vividly remember an occasion when, in the middle of a discussion of some philosophical
question, Radhakrishnan spontaneously recited a stanza from Nagarjuna's classic, *Fundamental Wisdom of the Middle Way*. It is most
remarkable that since Independence in 1947, India has maintained
the noble tradition of investing noted thinkers and scientists with
the nation's presidency.

After a difficult decade of adjustment, helping to settle the
community of around eighty thousand Tibetan refugees in various
parts of India, setting up schools for the youth, and attempting to
preserve the institutions of a threatened culture, I began my inter-
national travels toward the end of the 1960s. In addition to sharing
my understanding of the importance of basic human values, advocating interreligious understanding and harmony, and promoting
the rights and freedoms of the Tibetan people, I have taken the opportunity during my travels to meet distinguished scientists to discuss my interests, develop my knowledge, and delve ever deeper
into my understanding of science and its methods. Even as early as
the 1960s, I had discussed aspects of the interface between religion
and science with some valued visitors to my residence in Dharamsala, in north India. Two of the most memorable meetings of this
period were with the Trappist monk Thomas Merton, who had a

deep interest in Buddhism and opened my eyes to Christianity, and the scholar of religion Huston Smith.

One of my first teachers of science—and one of my closest scientific friends—was the German physicist and philosopher Carl von Weizsäcker, the brother of the West German president. Though he would describe himself as a politically active professor of philosophy who had been trained as a physicist, in the 1930s von Weizsäcker was employed as an assistant to the quantum physicist Werner Heisenberg. I will never forget von Weizsäcker's infectious and inspiring example as a man who constantly worried about the effects—especially the ethical and political consequences—of science. He sought relentlessly to apply the rigor of philosophical inquiry to the activity of science, in order to continually challenge it.

In addition to lengthy informal discussions on various occasions, I was fortunate to receive some formal tutorial sessions from von Weizsäcker on scientific topics. These were conducted in a style not so different from the one-to-one knowledge transmissions that are a familiar form of teaching in my own Tibetan Buddhist tradition. On more than one occasion, we were able to set aside two full days for a retreat when von Weizsäcker gave me an intensive tutorial on quantum physics and its philosophical implications. I feel deeply grateful for his tremendous kindness in granting me so much of his precious time and also for the depth of his patience, especially when I found myself struggling with a difficult concept, which I must admit was not infrequent.

Von Weizsäcker used to insist on the importance of empiricism in science. Matter can be known, he said, in two ways—it can be phenomenally given or it can be inferred. For instance, a brown spot on an apple can be seen with the eye; it is phenomenally given. But that there is a worm in the apple is something we may infer from the spot and from our general knowledge of apples and worms.

In Buddhist philosophy, there is the principle that the means by which a specific proposal is tested should accord with the nature of the subject under analysis. For example, if a proposal pertains to facts about the world that are observable, including one's own existence, then it is by empirical experience that the proposal may be affirmed or rejected. Thus, Buddhism puts the empirical method of direct observation first. If, by contrast, the proposal relates to generalizations that are induced from our experience of the world (for instance, the transient nature of life or the interconnectedness of reality), then it is by reason, primarily in the form of inference, that the proposal may be accepted or rejected. Thus, Buddhism accepts the method of reasoned inference—very much on the model presented by von Weizsäcker.

Finally, from the Buddhist point of view, there is a further level of reality, which may remain obscure to the unenlightened mind. Traditionally, a typical illustration of this would be the most subtle workings of the law of karma, and the question of why there are so many species of beings in the world. Only in this category of propositions is scripture cited as a potentially correct source of authority, on the specific basis that for Buddhists, the testimony of the Buddha has proven to be reliable in the examination of the nature of existence and the path to liberation. Although this principle of the three methods of verification—experience, inference, and a reliable authority—is implicit in the earliest developments of Buddhist thought, it was the great Indian logicians Dignaga (fifth century) and Dharmakirti (seventh century) who first formulated it as a systematic philosophical methodology.

In this final example, Buddhism and science clearly part company, since science, at least in principle, does not acknowledge any form of scriptural authority. But in the first two domains—the application of empirical experience and reason—there is a great methodological convergence between the two investigative traditions. In our day-to-day life, however, we regularly and habitually use the third method of testing claims about reality. For example, we accept the date of our birth on the verbal testimony of our relatives and in relation to the written testimony of a birth certificate. Even in science, we accept the results published by experimenters in peer-reviewed journals without ourselves repeating their experiments.

My engagement with science was undoubtedly given greater depth through my encounter with the remarkable physicist David Bohm, who had one of the greatest intellects and most open minds I have ever come across. I first met him in England in 1979, during my second trip to Europe, and we both felt an immediate rapport— indeed, I later found out that Bohm too had been an exile, having been forced to leave America during the persecutions of the McCarthy era. We began a lifelong friendship and a mutual intellectual exploration. David Bohm guided my understanding of the subtlest aspects of scientific thought, especially in physics, and exposed me to the scientific worldview at its best. While listening very carefully in a detailed conversation with a physicist like Bohm or von Weizsäcker, I would feel that I could grasp the intricacies of the full argument; unfortunately, when the sessions were over, there was often not a great deal left! My long discussion with Bohm over two decades fueled my own thinking about the ways Buddhist methods of inquiry may relate to those used in modern science.

I particularly admired Bohm's extraordinary openness to all areas of human experience, not only in the material world of his professional discipline but in all aspects of subjectivity, including the question of consciousness. In our conversations, I felt the presence of a great scientific mind which was prepared to acknowledge the value of observations and insights from other modes of knowledge than the objective scientific.

One of the particular qualities Bohm exemplified was the fascinating and essentially philosophical method of conducting a scientific inquiry by means of thought experiments. Simply put, this practice involves conjuring an imaginary scenario within which a specific hypothesis is tested by examining what consequences it may hold for assumptions normally thought to be irrefutable. A great deal of Einstein's work on the relativity of space and time was conducted by means of such thought experiments, which tested the understanding of physics that was current in his time. A famous example is the twins paradox, in which one brother remains on earth while another travels on a spaceship at a rate approaching the speed of light. For the brother on the ship, time should slow down. If he were to return ten years later, he would find that his brother on earth would have aged significantly more than he had. The full appreciation of this paradox requires an understanding of complex mathematical equations, which unfortunately lies beyond my skill.

In my engagement with science, I have always been extremely excited by this method of analysis because of its close parallels with Buddhist philosophical thinking. Before we met, Bohm had spent much time with the Indian spiritual thinker Jiddu Krishnamurti, and even participated in a number of dialogues with him.

On numerous occasions Bohm and I explored the ways objective scientific method may relate to meditative practice, which is, from the Buddhist point of view, equally empirical.

Although the basic emphases on empiricism and reason are similar in Buddhism and science, there are profound differences concerning what constitutes empirical experience and the forms of reasoning employed by the two systems. When Buddhism speaks of empirical experience, it has a broader understanding of empiricism, which includes meditative states as well as the evidence of the senses. Because of the development of technology in the last two hundred years, science has been able to extend the capacity of the senses to degrees unimaginable in earlier times. Hence scientists can use the naked eye, admittedly with the help of powerful instruments like microscopes and telescopes, to observe both remarkably minute phenomena, like cells and complex atomic structures, and the vast structures of the cosmos. On the basis of the expanded horizons of the senses, science has been able to push the limits of inference further than human knowledge has ever reached. Now, in response to traces left in bubble chambers, physicists can infer the existence of the constitutive particles of atoms, including even the elements within the neutron, such as quarks and gluons.

When I was a child experimenting with the telescope belonging to the Thirteenth Dalai Lama, I had a vivid experience of the power of inference based on empirical observation. In Tibetan folklore we speak of the rabbit on the moon—I believe the Europeans see a man instead of a rabbit. Anyway, one full-moon night in autumn, when the moon was especially clear, I decided to examine the rabbit with my telescope. To my surprise, I saw what looked like shadows. I was so excited that I insisted my two tutors come and peer through the telescope. I argued that the presence of shadows on the moon was proof that the moon is lit by the sun's light in the same way as the earth. They looked puzzled but agreed that the perception of shadows on the moon was indubitable. Later, when I saw photographs of lunar craters in a magazine, I noticed the same effect—that within the crater there was a shadow on one side but not on the other. From this I inferred that there must be a light source casting the shadow, just as on the earth. I concluded that the sun must be the source of the light that caused the shadows on the craters of the moon. I was very excited when I discovered later that this is in fact the case.

Strictly speaking, this process of reasoning is neither uniquely Buddhist nor uniquely scientific; rather it reflects a basic activity of the human mind which we naturally employ on a daily basis. The formal introduction to inference as a principle of logic for young trainee monks involves the illustration of how one may infer the presence of fire from a distance by seeing a column of smoke over a mountain pass, and from fire it would be normal in Tibet to infer human habitation. One can easily imagine a traveler, thirsty after a long day's walking, who feels the need for a cup of tea. He sees the smoke and thus infers fire and a dwelling where he can get shelter for the night. On the basis of this inference, the traveler is able to fulfill his desire to drink tea. From an observed phenomenon, directly evident to the senses, one can infer what remains hidden. This form of reasoning is common to Buddhism and science.

During my first visit to Europe, in 1973, I had the honor to encounter another of the twentieth century's great minds, the philosopher Sir Karl Popper. Like myself, Popper was once an exile—from his native Vienna during the period of Nazi rule—and he became one of the most articulate critics of totalitarianism. So we found much in common. Popper was an old man when I met him, over seventy, with bright eyes and great intellectual sharpness. I could guess how forceful he must have been in his youth from the passion he showed when we discussed the question of authoritarian regimes. In this meeting, Popper was more worried about the growing threat of communism, the perils of totalitarian political systems, the challenges of safeguarding individual liberty, and the sustenance of an open society than he was interested in exploring questions pertaining to the relations of science and religion. But we did discuss problems concerning method in science.

My English was not as good then as it is now, and my translators not so skilled. Unlike empirical science, philosophy and method are much more demanding to discuss. As a result, I benefited less perhaps from my opportunity to meet Popper than I did from meetings with figures like David Bohm and Carl von Weizsäcker. But we struck up a friendship, and I saw him again whenever I came to England, including a memorable visit in 1987 for tea at his house at Kenley in Surrey. I have a particular love of flowers and gardening, especially of orchids, and Sir Karl took great pride in giving me a tour of his own lovely garden and greenhouse. By this time I had discovered how great Popper's influence was in the philosophy of science, and especially on the question of scientific method.

One of Popper's principal contributions lay in clarifying the relative roles of inductive and deductive reasoning in the postulation and proof of scientific hypotheses. By induction, we mean building to a generalization from a series of empirically observed examples. Much of our everyday knowledge of the relations of cause and effect is inductive; for instance, on the basis of repeated observations of the correlation between smoke and fire, we make the generalization that where there is smoke there is fire. Deduction is the opposite process, operating from the knowledge of general truths to particular observations. For example, if one knows that all the cars produced in Europe after 1995 use only lead-free gas, when one hears that a particular car belonging to a friend was made in 2000, one can deduce that it must use lead-free gas. Of course, in science these forms are much more complex, especially deduction, because it involves the use of advanced mathematics.

One of the areas of reasoning where Buddhism and science differ lies in the role of deduction. What distinguishes science from Buddhism most in its application of reason is its highly developed use of profoundly complex mathematical reasoning. Buddhism, like all other classical Indian philosophies, has historically remained very concrete in its use of logic, whereby reasoning is never divorced from a particular context. In contrast, the mathematical reasoning of science allows a tremendous level of abstraction, so that the validity or invalidity of an argument can be determined purely on the basis of the correctness of an equation. Thus in one sense the generalization achievable through mathematics is at a much higher level than is possible in traditional forms of logic. Given the astounding success of mathematics, it is no wonder some people believe that the laws of mathematics are absolute and that mathematics is the true language of reality, intrinsic to nature herself.

Another of the differences between science and Buddhism as I see them lies in what constitutes a valid hypothesis. Here too Popper's delineation of the scope of a strictly scientific question represents a great insight. This is the Popperian falsifiability thesis, which states that any scientific theory must contain within it the conditions under which it may be shown to be false. For example, the theory that God created the world can never be a scientific one because it cannot contain an explanation of the conditions under which the theory could be proven false. If we take this criterion seriously, then many questions that pertain to our human existence, such as ethics, aesthetics, and spirituality, remain outside the domain of science. By contrast, the domain of inquiry in Buddhism is not limited to the objective. It also encompasses the subjective world of experience as well as the question of values. In other words, science deals with empirical facts but not with metaphysics and ethics, whereas for Buddhism, critical inquiry into all three is essential.

Popper's falsifiability thesis resonates with a major methodological principle in my own Tibetan Buddhist philosophical tradition. We might call this the "principle of the scope of negation." This principle states that there is a fundamental difference between that which is "not found" and that which is "found not to exist." If I look for something and fail to find it, this does not mean that the thing I am seeking does not exist. Not seeing a thing is not the same as seeing its non-existence. In order for there to be a coincidence between not seeing a thing and seeing its non-existence, the method of searching and the phenomenon being sought must be commensurate. For example, not seeing a scorpion on the page you are reading is adequate proof that there is no scorpion on the page. For if there were a scorpion on the page, it would be visible to the naked eye. However, not seeing acid in the paper on which the page is printed is not the same as seeing that the paper is acid-free, because to see acid in the paper one might require tools other than the naked eye. Additionally, the fourteenth-century philosopher Tsongkhapa argues there is a similar distinction between that which is negated by reason and that which is not affirmed by rea-

son, as well as between that which cannot withstand critical analysis and that which is undermined by such analysis.

These methodological distinctions may seem abstruse, but they
can have significant ramifications for one's understanding of the
scope of scientific analysis. For example, the fact that science has
not proven the existence of God does not mean that God does not
exist for those who practice in a theistic tradition. Likewise, just
because science has not proven beyond the shadow of a doubt that
beings take rebirth doesn't mean reincarnation isn't possible. In
science, the fact that we have not so far found life on any planet but
our own does not prove that life does not exist elsewhere.

So by the mid-1980s, on my numerous trips from India I had
met many scientists and philosophers of science and had participated in various conversations with them, both in public and in
private. Some of these, especially at the beginning, were not very
fruitful. In Moscow once, at the height of the Cold War, I had a
meeting with some scientists where my discussion of consciousness met with an immediate attack on the religious concept of the
soul, which they thought I was advocating. In Australia a scientist
opened his presentation with a somewhat hostile statement about
how he was there to defend science if there were to be any attack
on it from religion. However, 1987 marked an important stage in
my engagement with science. This was the year the first Mind and
Life conference took place at my residence in Dharamsala.

The meeting was organized by the Chilean neuroscientist
Francisco Varela, who taught in Paris, and the American business-
man Adam Engle. Varela and Engle approached me with a proposal
that they bring together a group of scientists from various disciplines who were sympathetic to the spirit of dialogue and that we
engage in an open-ended and informal private discussion that
would last for a week. I leapt at this idea. It was an extraordinary

opportunity to learn much more about science and to find out about the latest research and progress in scientific thinking. All the participants at the first meeting were so enthusiastic that the process has continued to this day, with a weeklong meeting every two years.

I first saw Varela at a conference in Austria. In the same year I had the opportunity to meet him one-to-one, and we started up an immediate friendship. Varela was a slim man with glasses and very soft-spoken. He combined a fine and logical mind with extraordinary clarity of expression, which made him an exceptional teacher. He took Buddhist philosophy and its contemplative tradition very seriously, but in his presentations he delivered the current thinking in the mainstream of science, unadorned and in an unbiased way. I cannot adequately say how grateful I am to Varela and Engle and also to Barry Hershey, who has generously provided the means to bring the scientists to Dharamsala. I have been assisted in these dialogues by two able interpreters, the American Buddhist scholar Alan Wallace and my translator Thupten Jinpa.

During this initial Mind and Life conference, I was exposed for the first time to a full historical account of the developments of scientific method in the West. Of particular interest to me was the idea of paradigm shifts—that is, fundamental changes within the worldview of a culture and their impacts on all aspects of scientific understanding. A classic example is the shift that occurred at the beginning of the twentieth century from classical Newtonian physics to relativity and quantum mechanics. Initially, the idea of paradigm shifts came as a shock to me. I had seen science as a relentless search for the ultimate truth about reality, in which new discoveries represented steps in humanity's collective incremental knowledge of the world. The ideal of this process would be that we might reach a final stage of complete and perfect knowledge. Now

I was hearing that there are subjective elements involved in the emergence of any particular paradigm and that there are therefore grounds for caution in speaking of a fully objective reality to which science gives us access.

When I speak with open-minded scientists and philosophers of science, it is clear that they have a deeply nuanced understanding of science and a recognition of the limits of scientific knowledge. At the same time, there are many people, both scientists and nonscientists, who appear to believe that all aspects of reality must and will fall within the scope of science. The assumption is sometimes made that, as society progresses, science will continually reveal the falsehoods of our beliefs—particularly religious beliefs—so that an enlightened secular society can eventually emerge. This is a view shared by Marxist dialectical materialists, as I discovered in my dealings with the leaders of Communist China in the 1950s and in the course of my studies of Marxist thought in Tibet. In this view, science is perceived as having disproved many of the claims of religion, such as the existence of God, grace, and the eternal soul. And within this conceptual framework, anything that is not proven or affirmed by science is somehow either false or insignificant. Such views are effectively philosophical assumptions that reflect their holders' metaphysical prejudices. Just as we must avoid dogmatism in science, we must ensure that spirituality is free from the same limitations.

Science deals with that aspect of reality and human experience that lends itself to a particular method of inquiry susceptible to empirical observation, quantification and measurement, repeatability, and intersubjective verification—more than one person has to be able to say, "Yes, I saw the same thing. I got the same results." So legitimate scientific study is limited to the physical world, including the human body, astronomical bodies, measurable energy, and how structures work. The empirical findings generated in this way form the basis for further experimentation and for generalizations that can be incorporated into the wider body of scientific knowledge. This is effectively the current paradigm of what constitutes science. Clearly, this paradigm does not and cannot exhaust all aspects of reality, in particular the nature of human existence. In addition to the objective world of matter, which science is masterful at exploring, there exists the subjective world of feelings, emotions, thoughts, and the values and spiritual aspirations based on them. If we treat this realm as though it had no constitutive role in our understanding of reality, we lose the richness of our own existence and our understanding cannot be comprehensive. Reality, including our own existence, is so much more complex than objective scientific materialism allows.

3
.
Emptiness,
Relativity, and
Quantum Physics

One of the most inspiring things about science is the change our understanding of the world undergoes in the light of new findings. The discipline of physics is still struggling with the implications of the paradigm shift it underwent as a result of the rise of relativity and quantum mechanics at the turn of the twentieth century. Scientists as well as philosophers have to live constantly with conflicting models of reality—the Newtonian model, assuming a mechanical and predictable universe, and relativity and quantum mechanics, assuming a more chaotic cosmos. The implications of the second model for our understanding of the world are still not entirely clear.

My own worldview is grounded in the philosophy and the teachings of Buddhism, which arose within the intellectual milieu of ancient India. I was exposed to ancient Indian philosophy at an early age. My teachers at that time were the then regent of Tibet,

Tadrak Rinpoche, and Ling Rinpoche. Tadrak Rinpoche was an old man, highly respected and quite stern. Ling Rinpoche was much younger; he was always gentle, soft-spoken, and deeply learned, but not a man of many words (at least when I was a child). I remember being terrified in the presence of both. I had several philosophical assistants to help me debate what I was taught. These included Trijang Rinpoche and the renowned Mongolian scholar-monk Ngodrup Tsoknyi. When Tadrak Rinpoche passed away, Ling Rinpoche became my senior tutor and Trijang Rinpoche was promoted to junior tutor.

These two remained my tutors until the end of my formal education, and I continually received numerous lineages of the Tibetan Buddhist heritage from both of them. They were close friends but very different characters. Ling Rinpoche was a stocky man with a shining bald head, and his whole body would shake when he laughed. He had a huge presence but small eyes. Trijang Rinpoche was a tall, thin man of great grace and elegance with a rather pointed nose for a Tibetan. He was gentle and had a deep voice, which was particularly melodious when he chanted. Ling Rinpoche was an acute philosopher with a sharp logical mind and a good debater with a phenomenal memory. Trijang Rinpoche was one of the greatest poets of his generation, with an eclectic command of art and literature. In terms of my own temperament and natural talents, I think I am closer to Ling Rinpoche than to any of my other tutors. It is perhaps fair to say that Ling Rinpoche has been the greatest influence on my life.

When I started learning about the various tenets of the ancient Indian schools, I had no way of relating these to any aspect of my personal experience. For example, the Samkhya theory of causation says that any effect is a manifestation of what was already there within the cause; the Vaisheshika theory of universals proposes that the plurality of any one class of objects has a permanent ideal generality that is independent of all the particulars. There were Indian theistic arguments proving the existence of the Creator and Buddhist counterarguments demonstrating the opposite. In addition, I had to learn many of the intricate differences in the tenets of the various schools within Buddhism. They were too esoteric to be of any immediate relevance to a boy in his early teens whose enthusiasm lay more in dismantling and reassembling watches or motorcars and poring over Second World War photographs in books and copies of *Life* magazine. In fact, when Babu Tashi took apart and cleaned the generator, I stood by to help him. I so enjoyed the process that I often forgot my studies and even my meals. When my philosophical assistants came to help me review, my mind would wander back to the generator and its many parts.

But things changed when I turned sixteen. Events were moving fast. As the Chinese army reached the border of Tibet in the summer of 1950, the regent, Tadrak Rinpoche, suggested that the time had come for me to assume full temporal leadership of the country. Perhaps it was this loss of youth, forced upon me by the grave reality of looming crisis, that brought home to me the true value of education. Whatever the cause may be, from the age of sixteen my engagement with the study of Buddhist philosophy, psychology, and spirituality was qualitatively different. Not only was I wholehearted in my pursuit of these studies but I also came to relate many aspects of what I studied to my own understanding of life and events in the world outside.

As I began to delve deeply into study, reflection, and meditative contemplation of Buddhist thought and practice, Tibet's dealings with the Chinese forces in the country—in the attempt to arrive at some kind of mutually satisfactory political arrangement—became more and more complicated. Finally, soon after I completed my formal education and sat in the holy city of Lhasa in the presence of several thousand monks for my Geshe examination, an event which marked the climax of my formal academic studies and which remains to this day a source of great satisfaction for me, the crisis in central Tibet forced me to escape from my homeland to India, to embark upon the life of a stateless refugee. This is still my legal status. But by losing citizenship in my own country, I found it in a larger sense; I can genuinely say that I am a citizen of the world.

One of the most important philosophical insights in Buddhism comes from what is known as the theory of emptiness. At its heart is the deep recognition that there is a fundamental disparity between the way we perceive the world, including our own existence in it, and the way things actually are. In our day-to-day experience, we tend to relate to the world and to ourselves as if these entities possess self-enclosed, definable, discrete, and enduring reality. For instance, if we examine our own conception of selfhood, we will find that we tend to believe in the presence of an essential core to our being, which characterizes our individuality and identity as a discrete ego, independent of the physical and mental elements that constitute our existence. The philosophy of emptiness reveals that this is not only a fundamental error but also the basis for attachment, clinging, and the development of our numerous prejudices.

According to the theory of emptiness, any belief in an objective reality grounded in the assumption of intrinsic, independent existence is untenable. All things and events, whether material, mental, or even abstract concepts like time, are devoid of objective, independent existence. To possess such independent, intrinsic existence would imply that things and events are somehow complete unto themselves and are therefore entirely self-contained. This would mean that nothing has the capacity to interact with and exert influence on other phenomena. But we know that there is cause and effect—turn a key in a starter, spark plugs ignite, the engine turns over, and gasoline and oil are burned. In a universe of self-contained, inherently existing things, these events would never occur. I would not be able to write on paper, and you would not be able to read the words on this page. So since we interact and change each other, we must assume that we are not independent—although we may feel or intuit that we are.

Effectively, the notion of intrinsic, independent existence is incompatible with causation. This is because causation implies contingency and dependence, while anything that possesses independent existence would be immutable and self-enclosed. Everything is composed of dependently related events, of continuously interacting phenomena with no fixed, immutable essence, which are themselves in constantly changing dynamic relations. Things and events are "empty" in that they do not possess any immutable essence, intrinsic reality, or absolute "being" that affords independence. This fundamental truth of "the way things really are" is described in the Buddhist writings as "emptiness," or *shunyata* in Sanskrit.

In our naïve or commonsense view of the world, we relate to things and events as if they possess an enduring intrinsic reality. We tend to believe that the world is composed of things and events, each of which has a discrete, independent reality of its own and it is these things with discrete identities and independence that interact with one another. We believe that intrinsically real seeds produce intrinsically real crops at an intrinsically real time in an intrinsically real place. Each member in this causal nexus—the seed, time, place, and effect—we take to have solid ontological status. This view of the world as made of solid objects and inherent properties is reinforced further by our language of subjects and predicates, which is structured with substantive nouns and adjectives on the one hand and active verbs on the other. But everything is constituted by parts—a person is body and mind both. Furthermore, the very identity of things is contingent upon many factors, such as the names we give them, their functions, and the concepts we have about them.

Although grounded in the interpretation of ancient scriptures which are attributed to the historical Buddha, this theory of emptiness was first systematically expounded by the great Buddhist philosopher Nagarjuna (c. second century C.E.). Little is known of his personal life, but he came from Southern India and he was— after the Buddha himself—the single most important figure for the formulation of Buddhism in India. Historians credit him with the emergence of the Middle Way school of Mahayana Buddhism, which remains the predominant school among Tibetans to this day. His most influential work in philosophy is *Fundamental Wisdom of the Middle Way,* which continues to be memorized, studied, and debated in the Tibetan monastic universities.

I spent much time in detailed study of the issues raised in this text, debating it with my teachers and colleagues. In the 1960s, during the first decade of my life as an exile in India, I was able to delve deeply and very personally into the philosophy of emptiness. Unlike today, my life then was reasonably relaxed, with relatively few formal engagements. I had not yet begun traveling the world, a process that now takes up a substantial part of my time. During this precious decade I had the fortune to spend many hours with my tutors, both of whom were experts in the philosophy and meditative practices of emptiness.

I also had teachings from a humble but gifted Tibetan scholar by the name of Nyima Gyaltsen. Affectionately known as Gen Nyima, he was one of those rare individuals with a gift for articulating profound philosophical insights in terms that are most accessible. He was slightly bald and wore large, round, tinted spectacles. He had an involuntary twitch in his right eye, which led to frequent blinking. But his powers of concentration, especially when following a complex train of thought and delving ever deeper into a point, were astounding—indeed, legendary. He could become totally oblivious to what was happening around him when he was in one of these states. The fact that the philosophy of emptiness was a particular specialty of Gen Nyima made my hours of exchange with him all the more rewarding.

One of the most extraordinary and exciting things about modern physics is the way the microscopic world of quantum mechanics challenges our commonsense understanding. The facts that light can be seen as either a particle or a wave, and that the uncertainty principle tells us we can never know at the same time what an electron does and where it is, and the quantum notion of superposition all suggest an entirely different way of understanding the world from that of classical physics, in which objects behave in a deterministic and predictable manner. For instance, in the well-known example of Schrödinger's cat, in which a cat is placed inside a box containing a radioactive source that has a 50 percent chance of releasing a deadly toxin, we are forced to accept that, until the lid is opened, this cat is both dead and alive, seemingly defying the law of contradiction.

To a Mahayana Buddhist exposed to Nagarjuna's thought, there is an unmistakable resonance between the notion of emptiness and the new physics. If on the quantum level, matter is revealed to be less solid and definable than it appears, then it seems to me that science is coming closer to the Buddhist contemplative insights of emptiness and interdependence. At a conference in New Delhi, I once heard Raja Ramanan, the physicist known to his colleagues as the Indian Sakharov, drawing parallels between Nagarjuna's philosophy of emptiness and quantum mechanics. After having talked to numerous scientist friends over the years, I have the conviction that the great discoveries in physics going back as far as Copernicus give rise to the insight that reality is not as it appears to us. When one puts the world under a serious lens of investigation—be it the scientific method and experiment or the Buddhist logic of emptiness or the contemplative method of meditative analysis—one finds things are more subtle than, and in some cases even contradict, the assumptions of our ordinary commonsense view of the world.

One may ask, Apart from misrepresenting reality, what is wrong with believing in the independent, intrinsic existence of things? For Nagarjuna, this belief has serious negative consequences. Nagarjuna argues that it is the belief in intrinsic existence that sustains the basis for a self-perpetuating dysfunction in our engagement with the world and with our fellow sentient beings. By according intrinsic properties of attractiveness, we react to certain objects and events with deluded attachment, while toward others, to which we accord intrinsic properties of unattractiveness, we react with deluded aversion. In other words, Nagarjuna argues that grasping at the independent existence of things leads to affliction, which in turn gives rise to a chain of destructive actions, reactions, and suffering. In the final analysis, for Nagarjuna, the theory of emptiness is not a question of the mere conceptual understanding of reality. It has profound psychological and ethical implications.

I once asked my physicist friend David Bohm this question: From the perspective of modern science, apart from the question of misrepresentation, what is wrong with the belief in the independent existence of things? His response was telling. He said that if we examine the various ideologies that tend to divide humanity, such as racism, extreme nationalism, and the Marxist class struggle, one of the key factors of their origin is the tendency to perceive things as inherently divided and disconnected. From this misconception springs the belief that each of these divisions is essentially independent and self-existent. Bohm's response, grounded in his work in quantum physics, echoes the ethical concern about harboring such beliefs that had worried Nagarjuna, who wrote nearly two thousand years before. Granted, strictly speaking, science does not deal with questions of ethics and value judgments, but the fact remains that science, being a human endeavor, is still connected to the basic question of the well-being of humanity. So in a sense, there is nothing surprising about Bohm's response. I wish there were more scientists with his understanding of the interconnectedness of science, its conceptual frameworks, and humanity.

As I understand it, modern science faced a crisis in the beginning of the twentieth century. The great edifice of classical physics developed by Isaac Newton, James Maxwell, and so many others, which provided such seemingly effective explanations for the perceived realities of the world and fitted so well with common sense, was undermined by the discovery of relativity and the strange behavior of matter at the subatomic level, which is explored in quantum mechanics. As Carl von Weizsäcker once explained it to me, classical physics accepted a mechanistic worldview in which certain universal physical laws, including gravity and the laws of mechanics, effectively determined the pattern of natural actions. In this model, there were four objective realities—bodies, forces, space, and time—and there was always a clear differentiation between the object as known and the subject who knows. But relativity and quantum mechanics, as von Weizsäcker put it, suggest that we must abolish as a matter of principle the separability of subject and object, and with this all our certainties about the objectifiability of our empirical data. Yet—and this is something von Weizsäcker insisted upon—the only terms we have for describing quantum mechanics and the experiments which verify its new picture of reality are those of classical physics, which quantum theory has disproved. Despite these problems, von Weizsäcker argued that we constantly need to search for the coherence in nature and for an understanding of reality, science, and the place of humanity that is more correct according to the latest scientific knowledge.

In the light of such scientific discoveries, I feel that Buddhism too must be willing to adapt the rudimentary physics of its early atomic theories, despite their long-established authority within the tradition. For example, the early Buddhist theory of atoms, which has not undergone any major revision, proposes that matter is constituted by a collection of eight so-called atomic substances: earth, water, fire, and air, which are the four elements, and form, smell, taste, and tactility, which are the four so-called derivative substances. The earth element sustains, water coheres, fire enhances, and air enables movement. An "atom" is seen as a composite of these eight substances, and on the basis of the aggregation of such composite "atoms," the existence of the objects in the macroscopic world is explained. According to one of the earliest Buddhist schools, Vaibhashika, these individual atomic substances are the smallest constituents of matter, indivisible and therefore without parts. When such "atoms" aggregate to form objects, Vaibhashika theorists assert that the individual atoms do not touch each other. Support from the air element and other forces in nature help the constitutive elements cohere into a system rather than collapsing inward or expanding indefinitely.

Needless to say, such theories must have developed through critical engagement with other Indian philosophical schools, especially the logical systems of Nyaya and Vaisheshika. If one examines Indian philosophical writings from antiquity, one senses a highly stimulating culture of debate, dialogue, and conversation between the adherents of different schools and systems. These classical Indian schools—such as Buddhism, Nyaya, Vaisheshika, Mimamsa, Samkhya, and Aidvaidavedanta—share basic interests and methods of analysis. This kind of intense debate between schools of thought has been a primary factor in the development of knowledge and the refinement of philosophical ideas, from the earliest period of Indian Buddhism to medieval and modern Tibet.

The earliest known sources of the Buddhist Vaibhashika theory of the atom may be Dharmashri's *Essence of Higher Knowledge* and the acclaimed *Great Treatise on Instantiation*. The first is generally placed by modern scholars sometime between the second century B.C.E. and the first century C.E. Although this work was never translated into Tibetan, I am told that a Chinese version was made sometime in the third century C.E. Dharmashri's text represents a sophisticated attempt to systematize the key standpoints of earlier Buddhist philosophy, so many of its basic ideas must have been current for some time before its composition. In contrast, the *Great Treatise* is a composite text, dated sometime between the first and third centuries C.E. The *Great Treatise* establishes the tenets of a specific Buddhist philosophical school as orthodox and responds to the various objections raised against these tenets by providing them with a rational philosophical foundation. Although the arguments of the *Great Treatise* are familiar to Tibetan Buddhism, the work itself was never in fact translated in full into Tibetan.

On the basis of these two texts, especially the latter, Vasubandhu, one of the great luminaries of Indian Buddhist philosophy, wrote his *Treasury of Higher Knowledge (Abhidharmakosha)* in the fourth century C.E. This summarizes key points from the *Great Treatise,* subjecting them to further analysis. It became a standard work on early Buddhist philosophy and psychology in Tibet; for example, as a young monk, I had to memorize the root text of Vasubandhu's *Treasury.*

As to the aggregation of atoms and the interrelation of atoms with their constituent substances, early Buddhism produced all kinds of speculative theories. It is interesting that in the *Treasury of Higher Knowledge* there is even a discussion of the physical size of different "atoms." The smallest utterly indivisible particle is said to be roughly 1-2,400th the size of a rabbit's "atom," whatever that means. I have no idea how Vasubandhu arrived at this calculation!

While accepting the basic atomic theory, other Buddhist schools questioned the notion of indivisible atoms. Some also queried the four derivative substances of form, smell, taste, and tactility as basic constituents of matter. For example, Vasubandhu himself is famous for his critique of the notion of objectively real, indivisible atoms. If independent, indivisible atoms exist, he argued, then it is impossible to account for the formation of the objects of the everyday world. In order for such objects to arise, there must be a way of explaining how simple atoms come together into complex composite systems.

If such aggregation takes place, as it must, then imagine a single atom surrounded by six different atoms, one in each of the four cardinal directions, one above, and one below. We can then ask: Does the same part of the central atom that touches the eastern atom also touch the atom in the north? If not, then the atom in the middle must have more than one part and is therefore at least conceptually divisible. The atom in the middle has a part that touches the atom in the east but does not touch the one in the north. If, by contrast, this eastern part does touch the atom in the north, there is nothing to prevent it from touching the atoms in all the remaining directions as well. In that case, Vasubandhu argues, the spatial locations of all the seven atoms—the one in the middle and the six surrounding it—will be the same, and all will collapse into a single atom. As a result of this thought experiment, Vasubandhu argues, it is impossible to account for the objects of the macroscopic world in terms of the aggregation of simple matter, such as indivisible atoms.

Personally, I have never understood the idea that qualities like smell, taste, and tactility are basic constituents of material objects. I can see how one could develop a coherent atomic theory of matter on the basis of the four elements as constituents. In any case, my own feeling is that this aspect of Buddhist thought, which is essentially a form of speculative, rudimentary physics, must now be modified in light of modern physics' detailed and experimentally verified understanding of the basic constituents of matter in terms of particles such as electrons revolving around a nucleus of protons and neutrons. When one listens to descriptions of subatomic particles, such as quarks and leptons, in modern physics, it is evident that the early Buddhist atomic theories and their conception of the smallest indivisible particles of matter are at best crude models. However, the basic thrust of the Buddhist theorists, that even the subtlest constituents of matter must be understood in terms of composites, appears to have been on the right track.

One of the principal motives underlying scientific and philosophical inquiry into the basic constituents of matter is to find matter's irreducible building block. This is true not only of ancient Indian philosophy and modern physics but also of the ancient Greek scientists, such as the "atomists." Effectively, this is a quest for the ultimate nature of reality, however one may define it. Buddhist thought argues on logical grounds that this search is misguided. At one stage science believed that in finding the atom it had found the ultimate constituent of matter, but twentieth-century experimental physics has subdivided the atom into ever more subtle particles. Although at least one view within quantum mechanics holds that we can never find an objectively real irreducible particle, many scientists still live in the hope of its discovery.

In the summer of 1998, I visited the Austrian physicist Anton Zeilinger's laboratory at the University of Innsbruck. Anton showed me an instrument that allows one to view an ionized single atom. Try as I might, though, I simply could not see it. Perhaps my karma wasn't ripe enough to enjoy this spectacle. I first met Anton when he came to a Mind and Life conference in Dharamsala in 1997. In some ways he is the opposite of David Bohm—a big man with a beard and glasses, a terrific sense of humor, and a full-bodied laugh. As an experimental physicist, he is remarkably open to any possible reformulation of the theoretical issues in light of the latest experimental results. His interest in a dialogue with Buddhism is in comparing theories of knowledge—quantum physics and Buddhism—because as he sees it, they both reject any notion of independent objective reality.

It was at this time too that I met the American physicist Arthur Zajonc. Arthur, who is quite soft-spoken and has piercing eyes, especially when he is dwelling deeply on a point, is a gifted teacher with the ability to make even the most complicated topics clear. As a moderator, Arthur would summarize and recapitulate arguments most succinctly, which was very helpful to me.

Several years before, I had been fortunate to visit the Niels Bohr Institute in Copenhagen to participate in an informal dialogue. A few days prior to this visit, during a brief stay in London, I had David Bohm and his wife to lunch in my hotel suite. Since I had told him that I was going to attend a dialogue on physics and Buddhist philosophy at the Bohr Institute, Bohm kindly brought me Niels Bohr's own two-page summary of his philosophical views on the nature of reality. It was fascinating to hear Bohm's account of Bohr's planetary model of the atom and Rutherford's model of the atom as a nucleus with orbiting electrons, both reactions to the "plum pudding" model.

The plum pudding model had arisen at the end of the nineteenth century, after J. J. Thomson discovered the negatively charged electron; it was assumed that the positive charge that balanced the electrons' negative charge was spread through the atom like a pudding, in which the electrons were plums. In the early twentieth century, Ernest Rutherford discovered that when positively charged alpha particles were fired at a gold film, most passed through but some bounced back. He correctly concluded that the positive charge of the gold atoms could not be spread through the atoms like a pudding but must be concentrated at their centers: when an alpha particle collided with the center of a gold atom, the positive charge was sufficient to repulse it. From this Rutherford formulated the "solar system" model of the atom, with a positively charged nucleus orbited by negative electrons. Niels Bohr was later to refine Rutherford's model with a planetary model of the atom that was in many ways the ancestor of quantum mechanics.

In our conversation Bohm also gave me a glimpse of the long-standing debate between Bohr and Einstein on the interpretation of quantum physics. The essence of this argument revolves around Einstein's refusal to accept the validity of the uncertainty principle; at the heart of the debate is the issue of whether reality at the fundamental level is indeterminate, unpredictable, and probabilistic. Einstein was deeply opposed to this possibility, as is reflected in his famous exclamation "God does not play dice!" All this reminded me of the history of my own Buddhist tradition, where debate has played a crucial role in the formation and refinement of many philosophical ideas.

Unlike the early Buddhist theorists, modern physicists can hugely enhance the power of their eyes through scientific instruments like giant telescopes, such as the Hubble telescope, or through electron microscopes. The result is an empirical knowledge of material objects that far surpasses even the imagination of ancient times. In view of this capability, I have made a strong case on several occasions for the introduction of basic physics into the studies of the Tibetan monastic colleges. I argued that we would not in fact be introducing a new subject; rather we would be updating an inherent part of the curriculum. I am happy that the academic monastic colleges now hold regular workshops on modern physics. These workshops are run by physics professors and some of their senior graduate students from Western universities. I hope that this initiative will ultimately result in the full entry of modern physics into the regular philosophical curriculum in Tibetan monasteries.

Although I had heard about Einstein's special theory of relativity a long time ago, it was again David Bohm who first explained it, along with some of its philosophical implications, to me. Because I have no mathematical background, teaching me modern physics, especially esoteric topics such as the theory of relativity, was not an easy task. When I think of Bohm's patience, his soft voice and gentle manner, and the care with which he made sure that I was following his explanation, I miss him dearly.

As any layperson who has attempted to understand this theory is aware, even a basic comprehension of Einstein's principle demands a willingness to defy common sense. Einstein put forward two postulates: the constancy of the speed of light, and his principle of relativity, which maintains that all laws of physics must be exactly the same for all observers in relative motion. With these two premises, Einstein revolutionized our scientific understanding of space and time.

His theory of relativity gave us the well-known equation of matter and energy, $E = mc^2$, admittedly the only scientific equation I know (today we can even see it on T-shirts), and a host of challenging and entertaining thought experiments. Many of these, such as the twins paradox of the special theory of relativity, time dilation, or the contraction of objects at high velocity, have now been confirmed experimentally. The twins paradox, in which if one twin were to fly aboard a spaceship at near the speed of light to a star say twenty light-years away and then return to earth, he will find his twin to be twenty years older than he is, reminds me of the story of how Asanga was taken to Maitreya's Heavenly Realm, where he received the five scriptures of Maitreya, a significant set of Mahayana texts, all in the time frame of a tea break. But when he returned to earth, fifty years had passed.

Appreciating the full nature of the twins paradox involves understanding a set of complex calculations which I am afraid are beyond me. As I understand it, the most important implication of Einstein's theory of relativity is that notions of space, time, and mass cannot be seen as absolutes, existing in themselves as permanent, unchanging substances or entities. Space is not an independent, three-dimensional domain, and time is not a separate entity; rather they coexist as a four-dimensional continuum of "space-time." In a nutshell, Einstein's special theory of relativity implies that, while the speed of light is invariable, there is no absolute, privileged frame of reference, and everything, including space and time, is ultimately relative. This is a truly remarkable revelation.

In the Buddhist philosophical world, the concept of time as relative is not alien. Before the second century C.E., the Sautrantika school argued against the notion of time as absolute. Dividing the temporal process into the past, present, and future, the Sautrantikas demonstrated the interdependence of the three and argued for the untenability of any notion of independently real past, present, and future. They showed that time cannot be conceived as an intrinsically real entity existing independently of temporal phenomena but must be understood as a set of relations among temporal phenomena. Apart from the temporal phenomena upon which we construct the concept of time, there is no real time that is somehow the grand vessel in which things and events occur, an absolute that has an existence of its own.

These arguments for the relativity of time, subsequently developed by Nagarjuna, are primarily philosophical, but the fact remains that time has been perceived as relative in the Buddhist philosophical tradition for nearly two thousand years. Although I am told that some scientists view Einstein's four-dimensional space-time as a grand vessel with inherent existence in which events occur, for a Buddhist thinker familiar with Nagarjuna's arguments, Einstein's demonstration of the relativity of time, especially through his famous thought experiments, is extremely helpful in deepening the understanding of the relative nature of time.

My grasp of quantum theory is, I confess, not all that good—

though I have tried very hard! I am told that one of the greatest of all quantum theorists, Richard Feynman, wrote, "I think I can safely say that nobody understands quantum mechanics," so at least I feel I am in good company. But even for someone like myself who cannot follow the complex mathematical details of the theory—in fact, mathematics is one area of modern science to which I seem to have no karmic connection at all—it is apparent that we cannot speak of subatomic particles as determinate, independent, or mutually exclusive entities. Elementary constituents of matter and photons (which is to say, the basic substances, respectively, of matter and of light) can be either particles or waves or both. (In fact, the man who won the Nobel Prize for showing that the electron is a wave, George Thomson, was the son of the man who won the same prize for showing that the electron is a particle, J. J. Thomson.) Whether one perceives electrons as particles or waves, I am told, is dependent on the action of the observer and his or her choice of apparatus or measurement.

Although I had long heard of this paradoxical nature of light, only in 1997—when the experimental physicist Anton Zeilinger explained it to me with detailed illustrations—did I feel I had finally managed to grasp the issue. Anton showed how it is the experiment itself that determines whether an electron behaves as a particle or as a wave. In the famous double-slit experiment, electrons are fired one at a time through an interference barrier with two slits and are registered on material such as a photographic plate behind the barrier. If one slit is open, each electron makes an imprint on the photographic plate in the manner of a particle. However, if both slits are open, when a large number of electrons are fired, the imprint left on the photographic plate indicates that they have passed through both slits at the same time, leaving a wavelike pattern.

Anton brought an apparatus that could repeat this experiment on a smaller scale, so all the participants had great fun. Anton likes to remain very close to the empirical aspects of quantum mechanics, grounding all his understanding in what we can directly learn from experiments. This was quite a different approach from that of David Bohm, who was primarily interested in the theoretical and philosophical implications of quantum mechanics. I later learned that Anton was and remains a strong advocate of what is called the Copenhagen interpretation of quantum mechanics, while David Bohm was one of its strongest critics.

I must admit I am still not quite sure what the full conceptual and philosophical implications of this paradox of wave-particle duality might be. I have no problem in accepting the basic philosophical implication, that at the subatomic level the very notion of reality cannot be divorced from the system of measurements used by an observer, and cannot therefore be said to be completely objective. However, this paradox also seems to suggest that—unless one accords some kind of intelligence to electrons—at the subatomic level two of the most important principles of logic, the law of contradiction and the law of the excluded middle, appear to break down. In normal experience, we would expect that what is a wave cannot be a particle, yet at the quantum level, light appears to be a contradiction because it behaves as both. Similarly, in the double-slit experiment, it appears that some of the photons pass through both slits at the same time, thus breaking the law of the excluded middle, which expects them to pass through either one slit or the other.

Regarding the conceptual implications of the results of the double-slit experiment, I think there is still considerable debate. Heisenberg's famous uncertainty principle states that the more precise one's measurement of an electron's position the more uncertain is one's knowledge of its momentum, and the more precise one's measurement of its momentum the more uncertain one is of its position. One can know at any one time where an electron is but not what it is doing, or what it is doing but not where it is. Again this shows that the observer is fundamental: in choosing to learn an electron's momentum, we exclude learning its position; in choosing to learn its position, we exclude learning its momentum. The observer, then, is effectively a participant in the reality being observed. I realize that this issue of the observer's role is one of the thorniest questions in quantum mechanics. Indeed, at the Mind and Life conference in 1997, the various scientific participants held differently nuanced views. Some would argue that the observer's role is limited to the choice of measuring apparatus, while others accord greater importance to the observer's role as a constitutive element in the reality being observed.

This issue has long been a focus of discussion in Buddhist thought. On one extreme are the Buddhist "realists," who believe that the material world is composed of indivisible particles which have an objective reality independent of the mind. On the other extreme are the "idealists," the so-called Mind-only school, who reject any degree of objective reality in the external world. They perceive the external material world to be, in the final analysis, an extension of the observing mind. There is, however, a third standpoint, which is the position of the Prasangika school, a perspective held in the highest esteem by the Tibetan tradition. In this view, although the reality of the external world is not denied, it is understood to be relative. It is contingent upon our language, social conventions, and shared concepts. The notion of a pre-given, observer-independent reality is untenable. As in the new physics, matter cannot be objectively perceived or described apart from the observer—matter and mind are co-dependent.

This recognition of the fundamentally dependent nature of reality—called "dependent origination" in Buddhism—lies at the very heart of the Buddhist understanding of the world and the nature of our human existence. In brief, the principle of dependent origination can be understood in the following three ways. First, all conditioned things and events in the world come into being only as a result of the interaction of causes and conditions. They don't just arise from nowhere, fully formed. Second, there is mutual dependence between parts and the whole; without parts there can be no whole, without a whole it makes no sense to speak of parts. This interdependence of parts and the whole applies in both spatial and temporal terms. Third, anything that exists and has an identity does so only within the total network of everything that has a possible or potential relation to it. No phenomenon exists with an independent or intrinsic identity.

And the world is made up of a network of complex interrelations. We cannot speak of the reality of a discrete entity outside the context of its range of interrelations with its environment and other phenomena, including language, concepts, and other conventions. Thus, there are no subjects without the objects by which they are defined, there are no objects without subjects to apprehend them, there are no doers without things done. There is no chair without legs, a seat, a back, wood, nails, the floor on which it rests, the walls that define the room it's in, the people who constructed it, and the individuals who agree to call it a chair and recognize it as something to sit on. Not only is the existence of things and events utterly contingent but, according to this principle, their very identities are thoroughly dependent upon others.

In physics, the deeply interdependent nature of reality has been brought into sharp focus by the so-called EPR paradox—named after its creators, Albert Einstein, Boris Podolsky, and Nathan Rosen—which was originally formulated to challenge quantum mechanics. Say a pair of particles is created and then separates, moving away from each other in opposite directions—perhaps to greatly distant locations, for example, Dharamsala, where I live, and say, New York. One of the properties of this pair of particles is that their spin must be in opposite directions—so that one is measured as "up" and the other will be found to be "down." According to quantum mechanics, the correlation of measurements (for example, when one is up, then the other is down) must exist even though the individual attributes are not determined until the experimenters measure one of the particles, let us say in New York. At that point, the one in New York will acquire a value—let us say up—in which case the other particle must simultaneously become down. These determinations of up and down are instantaneous, even for the particle at Dharamsala, which has not itself been measured. Despite their separation, the two particles appear as an entangled entity. There seems, according to quantum mechanics, to be a startling and profound interconnectedness at the heart of physics.

Once at a public talk in Germany, I drew attention to the growing trend among serious scientists of taking the insights of the world's contemplative traditions into account. I spoke about the meeting ground between my own Buddhist tradition and modern science—especially in the Buddhist arguments for the relativity of time and for rejecting any notion of essentialism. Then I noticed von Weizsäcker in the audience, and when I described my debt to him for what little understanding of quantum physics I possess, he graciously commented that if his own teacher Werner Heisenberg had been present, he would have been excited to hear of the clear, resonant parallels between Buddhist philosophy and his scientific insights.

Another significant set of issues in quantum mechanics concerns the question of measurement. I gather that, in fact, there is an entire area of research dedicated to this matter. Many scientists say that the act of measurement causes the "collapse" of either the wave or the particle function, depending upon the system of measurement used in the experiment; only upon measurement does the potential become actual. Yet we live in a world of everyday objects. So the question is, How, from the point of view of physics, do we reconcile our commonsense notions of an everyday world of objects and their properties on the one hand and the bizarre world of quantum mechanics on the other? Can these two perspectives be reconciled at all? Are we condemned to live with what is apparently a schizophrenic view of the world?

At a two-day retreat on the epistemological issues pertaining to the foundations of quantum mechanics and Buddhist Middle Way philosophy at Innsbruck, where Anton Zeilinger, Arthur Zajonc, and I met for a dialogue, Anton told me that a well-known colleague of his once remarked that most quantum physicists relate to their field in a schizophrenic manner. When they are in the laboratory and play around with things, they are realists. They talk about photons and electrons going here and there. However, the moment you switch into philosophical discussion and ask them about the foundation of quantum mechanics, most would say that nothing really exists without the apparatus defining it.

Somewhat parallel problems arose in Buddhist philosophy in relation to the disparity between our commonsense view of the world and the perspective suggested by Nagarjuna's philosophy of emptiness. Nagarjuna invoked the notion of two truths, the "conventional" and the "ultimate," relating respectively to the everyday world of experience and to things and events in their ultimate mode of being, that is, on the level of emptiness. On the conventional level, we can speak of a pluralistic world of things and events with distinct identities and causation. This is the realm where we can also expect the laws of cause and effect, and the laws of logic—such as the principles of identity, contradiction, and the law of the excluded middle—to operate without violation. This world of empirical experience is not an illusion, nor is it unreal. It is real in that we experience it. A grain of barley does produce a barley sprout, which can eventually yield a barley crop. Taking a poison can cause one's death and, similarly, taking a medication can cure an illness. However, from the perspective of the ultimate truth, things and events do not possess discrete, independent realities. Their ultimate ontological status is "empty" in that nothing possesses any kind of essence or intrinsic being.

I can envision something similar to this principle of two truths applying in physics. For instance, we can say that the Newtonian model is an excellent one for the commonsense world as we know it, while Einsteinian relativity—based on radically different presuppositions—represents in addition an excellent model for a different or more inclusive domain. The Einsteinian model describes aspects of reality for which the states of relative motion are crucial but does not really affect our commonsense picture under most circumstances. Likewise, the quantum physics models of reality represent the workings of a different domain—the mostly "inferred" reality of particles, especially in the arena of the microscopic. Each of these pictures is excellent in its own right and for the purposes for which it was designed, but if we believe any of these models to be constituted by intrinsically real things, we are bound to be disappointed.

Here I find it helpful to reflect on a critical distinction drawn by Chandrakirti (seventh century C.E.) in relation to the domains of discourse that pertain to the conventional and the ultimate truths of things. Chandrakirti argues that, when formulating one's understanding of reality, one must be sensitive to the scope and parameters of the specific mode of inquiry. For example, he argues that to reject distinct identity, causation, and origination within the everyday world, as some interpreters of the philosophy of emptiness had suggested, simply because these notions are untenable from the perspective of ultimate reality, constitutes a methodological error.

On a conventional level, we see cause and effect all the time. When we're trying to find who's at fault in an accident, we are not delving into the deeper nature of reality, where an infinite chain of events would make it impossible to place blame. When we accord such characteristics as cause and effect to the empirical world, we are not working on the basis of a metaphysical analysis that probes the ultimate ontological status of things and their properties. We do so within the boundaries of everyday convention, language, and logic. In contrast, Chandrakirti argues, the metaphysical postulates of philosophical schools, such as the concept of the Creator or the eternal soul, can be negated through the analysis of their ultimate ontological status. This is because these entities are posited on the basis of an exploration into the ultimate mode of being of things.

In essence, Nagarjuna and Chandrakirti are suggesting this: when we relate to the empirical world of experience, so long as we do not invest things with independent, intrinsic existence, notions of causation, identity, and difference, and the principles of logic will continue to remain tenable. However, their validity is limited to the relative framework of conventional truth. Seeking to ground notions such as identity, existence, and causation in an objective, independent existence is transgressing the bounds of logic, language, and convention. We do not need to postulate the objective, independent existence of things, since we can accord robust, nonarbitrary reality to things and events that not only support everyday functions but also provide a firm basis for ethics and spiritual activity. The world, according to the philosophy of emptiness, is constituted by a web of dependently originating and interconnected realities, within which dependently originated causes give rise to dependently originated consequences according to dependently originating laws of causality. What we do and think in our own lives, then, becomes of extreme importance as it affects everything we're connected to.

The paradoxical nature of reality revealed in both the Buddhist philosophy of emptiness and modern physics represents a profound challenge to the limits of human knowledge. The essence of the problem is epistemological: How do we conceptualize and understand reality coherently? Not only have Buddhist philosophers of emptiness developed an entire understanding of the world based on the rejection of the deeply ingrained temptation to treat reality as if it were composed of intrinsically real objective entities but they have also striven to live these insights in their day-to-day lives. The Buddhist solution to this seeming epistemological contradiction involves understanding reality in terms of the theory of two truths. Physics needs to develop an epistemology that will help resolve the seemingly unbridgeable gulf between the picture of reality in classical physics and everyday experience and that in their quantum mechanics counterpart. As for what an application of the two truths in physics might look like, I simply have no idea. At its root, the philosophical problem confronting physics in the wake of quantum mechanics is whether the very notion of reality—defined in terms of essentially real constituents of matter—is tenable. What the Buddhist philosophy of emptiness can offer is a coherent model of understanding reality that is non-essentialist. Whether this could prove useful only time will tell.

4.

THE BIG BANG AND
THE BUDDHIST
BEGINNINGLESS
UNIVERSE

Who has not felt a sense of awe while looking deep into the skies lit with countless stars on a clear night? Who has failed to wonder whether there is an intelligence behind the cosmos? Who has not asked themselves if ours is the only planet to support living creatures? To me, these are natural curiosities in the human mind. Throughout the history of human civilization, there has been a real impulse to find answers to these questions. One of the great achievements of modern science is that it seems to have brought us closer than ever to an understanding of the conditions and complicated processes underlying the origins of our cosmos.

Like many ancient cultures, Tibet has a complex system of astrology that contains elements of what modern culture would call astronomy, so there are Tibetan names for most of the stars visible to the naked eye. In fact, Tibetans and Indians have long been able to predict lunar and solar eclipses with a high degree of accuracy on the basis of their astronomical observations. As a child in Tibet, I spent many nights peering into the sky with my telescope, learning the shapes and names of the constellations.

I remember to this day the joy I felt when I was able to visit a real astronomical observatory in Delhi at the Birla Planetarium. In 1973, during my first visit to the West, I was invited by Cambridge University in England to give a talk at the Senate House and the Faculty of Divinity. When the vice chancellor asked me if there was something I especially wanted to do in Cambridge, I responded without hesitation that I wished to visit the famous radio telescope at the Department of Astronomy.

At one of the Mind and Life conferences in Dharamsala, the astrophysicist Piet Hut, from the Institute for Advanced Study in Princeton, showed a computer simulation of how astronomers envision cosmic events unfolding when galaxies collide. It was a fascinating sight, a real spectacle. Such computer animations help one visualize the way, given certain conditions immediately after the cosmic explosion, the universe unfolded through time following the basic laws of cosmology. After Piet Hut's presentation, we had an open discussion. Two of the other participants at the meeting, David Finkelstein and George Greenstein, tried to demonstrate the phenomenon of the expanding universe by using elastic bands with rings on them. I remember this clearly because my two translators and I had some difficulty visualizing cosmic expansion from this demonstration. Later, all the scientists at the meeting joined in to try to simplify the explanation, which of course had the effect of confusing us even more.

Modern cosmology—like so much else in the physical sciences—is founded on Einstein's theory of relativity. In cosmology, astronomical observations taken together with the theory of general relativity, which reformulated gravity as the curvature of both space and time, have shown that our universe is neither eternal nor static in its current form. It is continuously evolving and expanding. This finding accords with the basic intuition of the ancient Buddhist cosmologists, who conceived that any particular universe system goes through stages of formation, expansion, and ultimately destruction. In modern cosmology, in the 1920s, both theoretical prediction (by Alexander Friedmann) and detailed empirical observation (by Edwin Hubble)—for instance, the observation that a larger red shift is measurable in the light emitted by distant galaxies than in that emitted by nearer ones—demonstrated convincingly that the universe is curved and expanding.

It is assumed that this expansion emerged from a great cosmic explosion—the famous big bang, which is thought to have occurred 12 to 15 billion years ago. Most cosmologists today believe that a few seconds after this explosion, the temperature decreased to a point where reactions occurred that began making the nuclei of the lighter elements, from which much later all the matter in the cosmos came into being. Thus all of space, time, matter, and energy as we know and experience them came into being from this fireball of matter and radiation. In the 1960s background microwave radiation was detected throughout the universe; it came to be recognized as an echo, or afterglow, of the events of the big bang. Precise measurement of the spectrum, polarization, and spatial distribution of this background radiation has apparently confirmed, at least in outline, current theoretical models of the origins of the universe.

Until the accidental detection of this background microwave noise, there was an ongoing debate between two powerful schools in modern cosmology. Some preferred to understand the expansion of the universe as a steady state theory, meaning that the universe is expanding at a steady rate, with constant laws of physics

applying at all times. On the other side were those who saw evolution in terms of a cosmic explosion. I am told that the proponents
of the steady state model included some of the greatest minds of
modern cosmology, such as Fred Hoyle. In fact, at one point within
living memory, this theory was the mainstream scientific view of
the origin of our universe. Today, it seems, most cosmologists are
convinced that the background microwave noise conclusively
demonstrates the validity of the big bang hypothesis. This is a wonderful example of how in science, in the final analysis, it is empirical evidence that represents the last court of justice. At least in
principle, this is true also in Buddhist thought, where it is said that
to defy the authority of empirical evidence is to disqualify oneself
as someone worthy of critical engagement in a dialogue.

In Tibet there were complex myths of creation that originated
in the pre-Buddhist religion of Bön. A central theme in these
myths is the bringing of order out of chaos, light out of darkness,
day out of night, existence out of nothingness. These acts are effected by a transcendent being, who creates everything out of pure
potentiality. Another set of myths portrays the universe as a living
organism born from a cosmic egg. In the rich spiritual and philosophical traditions of ancient India, numerous conflicting cosmological views were developed. These included such diverse
formulations as the early Samkhya theory of primordial materiality, which describes the origins of the cosmos and life within it as
an expression of an underlying absolute substratum; Vaisheshika
atomism, which substituted a plurality of indivisible "atoms" as the
basic units of reality for a single underlying substratum; the various theories of the gods Brahman or Ishvara as the source of divine creation; and the radical materialist Charvaka school's theory
of the evolution of the universe through a purposeless, random development of matter, with all mental processes viewed as derivative

of complex configurations of material phenomena. This last position is not dissimilar to scientific materialism's belief that mind is reducible to neurological and biochemical reality and these in turn to facts of physics. Buddhism, by contrast, explains the evolution of the cosmos in terms of the principle of dependent origination, in that the origin and existence of everything has to be understood in terms of the complex network of interconnected causes and conditions. This applies to consciousness as well as matter.

According to the early scriptures, the Buddha himself never directly answered questions put to him about the origin of the universe. In a famous simile, the Buddha referred to the person who asks such questions as a man wounded by a poisoned arrow. Instead of letting the surgeon pull the arrow out, the injured man insists first on discovering the caste, name, and clan of the man who shot the arrow; whether he is dark, brown, or fair; whether he lives in a village, town, or city; whether the bow used was a longbow or a crossbow; whether the bowstring was fiber, reed, hemp, sinew, or bark; whether the arrow shaft was of wild or cultivated wood; and so forth. Interpretations of the meaning of the Buddha's refusal to answer these questions directly vary. One view is that the Buddha refused to answer because these metaphysical questions do not directly pertain to liberation. Another view, primarily argued by Nagarjuna, is that insofar as the questions were framed on the presupposition of the intrinsic reality of things, and not on dependent origination, responding would have led to a deeper entrenchment in the belief in solid, inherent existence.

The questions are grouped slightly differently in the different Buddhist traditions. The Pali canon lists ten such "unanswered" questions, while the classical Indian tradition inherited by the Tibetans lists the following fourteen:

1. Are the self and the universe eternal?

2. Are the self and the universe transient?

3. Are the self and the universe both eternal and transient?

4. Are the self and the universe neither eternal nor transient?

5. Do the self and the universe have a beginning?

6. Do the self and the universe have no beginning?

7. Do the self and the universe have both beginning and no beginning?

8. Do the self and the universe have neither beginning nor no beginning?

9. Does the Blessed One exist after death?

10. Does the Blessed One not exist after death?

11. Does the Blessed One both exist and not exist after death?

12. Does the Blessed One neither exist nor not exist after death?

13. Is the mind the same as the body?

14. Are the mind and body two separate entities?

Despite the scriptural tradition of the Buddha's refusal to engage on this level of metaphysical discourse, Buddhism as a philosophical system in ancient India developed a long history of delving deeply into these fundamental and perennial questions about our existence and the world we live in. My own Tibetan tradition has inherited this philosophical legacy.

There were two main traditions of cosmology in Buddhism. One is the Abhidharma system, shared by many Buddhist schools, such as Theravada Buddhism, which is the dominant tradition to this day in countries like Thailand, Sri Lanka, Burma, Cambodia, and Laos. Although the tradition of Buddhism that came to Tibet is Mahayana Buddhism, especially the version of Indian Buddhism

known as the Nalanda tradition, Abhidharma psychology and cosmology became an important part of the Tibetan intellectual land-
scape. The primary work on the Abhidharma system of cosmology
that made its way into Tibet was Vasubandhu's *Treasury of Higher
Knowledge* (*Abhidharmakosha*). The second cosmological tradition
in Tibet is the system found in a collection of important Vajrayana
Buddhist texts belonging to the genre of theory and practice
known as the Kalachakra, which literally means "wheel of time." Although tradition attributes the core teachings of the Kalachakra cycle to the Buddha, it is difficult to identify precisely the date of
origin of the earliest known works in this system. Following the
translation of the key Kalachakra texts from Sanskrit into Tibetan
in the eleventh century, Kalachakra came to occupy an important
place in the Tibetan Buddhist heritage.

By the age of twenty, when I began my systematic study of the
texts that discuss Abhidharma cosmology, I knew that the world was
round, had looked at photographic images of volcanic craters on
the surface of the moon in magazines, and had some inkling of the
orbital rotation of earth and moon around the sun. So I must admit,
when I was studying Vasubandhu's classic presentation of the Abhidharma cosmological system, it did not much appeal to me.

Abhidharma cosmology describes a flat earth, around which
celestial bodies like the sun and moon revolve. According to this
theory, our earth is one of the four "continents"—in fact, the southern continent—which lie in the four cardinal directions of a tow-
ering mountain called Mount Meru, at the center of the universe.
Each of these continents is flanked by two smaller continents,
while the gaps between them are filled with massive oceans. This
entire world system is supported by a "ground," which in turn remains suspended in empty space. The power of "air" keeps the base
afloat in empty space. Vasubandhu gives a detailed description of

the orbital passage of sun and moon, their sizes, and their distances from the earth.

These sizes, distances, and so forth are flatly contradicted by the empirical evidence of modern astronomy. There is a dictum in Buddhist philosophy that to uphold a tenet that contradicts reason is to undermine one's credibility; to contradict empirical evidence is a still greater fallacy. So it is hard to take the Abhidharma cosmology literally. Indeed, even without recourse to modern science, there is a sufficient range of contradictory models for cosmology within Buddhist thought for one to question the literal truth of any particular version. My own view is that Buddhism must abandon many aspects of the Abhidharma cosmology.

To what extent Vasubandhu himself believed in the Abhidharma worldview is open to question. He was presenting systematically the variety of cosmological speculations that were then current in India. Strictly speaking, the description of the cosmos and its origins—which the Buddhist texts refer to as "the container"—is secondary to the account of the nature and origins of sentient beings, who are "the contained." The Tibetan scholar Gendün Chöphel, who traveled extensively through the Indian subcontinent in the 1930s, suggested that the Abhidharma description of the "earth" as the southern continent represents an ancient map of central India. He gave a compelling account of how the descriptions of the three remaining "continents" match actual geographical sites in modern India. Whether this hunch is true or whether these places were in fact named after the "continents" thought to surround Mount Meru remains an open question.

In some early scriptures, the planets are described as spherical bodies suspended in empty space, not dissimilar to the conception of planetary systems in modern cosmology. In Kalachakra cosmology, a definite sequence is given for the evolution of the celestial bodies in our present galaxy. First, the stars were formed, after which the solar system came into being, and so on. What is interesting in both the Abhidharma and the Kalachakra cosmologies is the big picture they offer of the origin of the universe. There is a recognition that ours is but one among countless world systems. Both the Abhidharma and Kalachakra give the technical term *trichilicosm* (which I believe corresponds roughly to a billionfold world system) to convey this notion of vast universe systems, and both claim that there are countless such systems. So in principle, although there is no "beginning" or "end" to the universe as a whole, there is a definite temporal process of a beginning, middle, and end in relation to any individual world system.

The evolution of a particular universe system is understood in terms of four principal stages, known as the eras of (1) emptiness, (2) formation, (3) abiding, and finally, (4) destruction. Each of these stages is thought to last a tremendously long time, twenty "medium aeons," and it is only in the last medium aeon of the formation stage that sentient beings are said to evolve. The destruction of a universe system may be caused by any of the three natural elements other than earth and space—namely, water, fire, and air. Whichever element led to the destruction of the previous world system will act as the basis for the creation of a new universe.

At the heart of Buddhist cosmology is, therefore, not only the idea that there are multiple world systems—infinitely more than the grains of sand in the river Ganges, according to some texts—but also the idea that they are in a constant state of coming into being and passing away. This means that the universe has no absolute beginning. The questions this idea poses for science are fundamental. Was there one big bang or were there many? Is there one universe or are there many, or even an infinite number? Is the universe finite or infinite, as the Buddhists assert? Will our universe expand infinitely, or will its expansion slow down, even reverse, so that it ultimately ends in a big crunch? Is our universe part of an eternally reproducing cosmos? Scientists are debating these issues intensely. From the Buddhist point of view, there is this further question. Even if we grant that there was only one big cosmic bang, we can still ask, Is this the origin of the entire universe or does this mark only the origin of our particular universe system? So a key question is whether the big bang—which, according to modern cosmology, is the beginning of our current world system—is really the beginning of everything.

From the Buddhist perspective, the idea that there is a single definite beginning is highly problematic. If there were such an absolute beginning, logically speaking, this leaves only two options. One is theism, which proposes that the universe is created by an intelligence that is totally transcendent, and therefore outside the laws of cause and effect. The second option is that the universe came into being from no cause at all. Buddhism rejects both these options. If the universe is created by a prior intelligence, the questions of the ontological status of such an intelligence and what kind of reality it is remain.

The great logician and epistemologist Dharmakirti (seventh century C.E.) cogently presented the standard Buddhist critique of theism. In his classic *Exposition of Valid Cognition*, Dharmakirti takes to task some of the most influential "proofs" for the existence of the Creator formulated by the theistic Indian philosophical schools. Briefly put, the arguments for theism run as follows: The worlds of both inner experience and external matter are created by a preceding intelligence, (a) because, like carpenter's tools, they operate in a successive sequence of orderliness; (b) like artifacts such as vases, they have forms; and (c) like the objects of everyday use, they possess causal efficacy.

These arguments, I think, have a resemblance to a theistic argument in the Western philosophical tradition known as the argument from design. This argument takes the high degree of order that we perceive in nature as evidence of an intelligence which must have brought it into being. Just as one cannot conceive of a watch without a watchmaker, so it is difficult to conceive of an orderly universe without an intelligence behind it.

The classical Indian philosophical schools that espouse a theistic understanding of the origin of the universe are as diverse as their counterparts in the West. One of the earliest is a branch of the Samkhya school, which upheld the view that the universe came into being through the creative interplay of what they call "the primal substance," *prakrit*, and *Ishvara*, God. This is a sophisticated metaphysical theory grounded in the natural law of causality, explaining the Divinity's role in terms of the more mysterious features of reality, such as creation, the purpose of existence, and other such matters.

The crux of Dharmakirti's critique involves demonstrating a fundamental inconsistency he perceives in the theistic standpoint. He shows that the very endeavor of accounting for the origin of the universe in theistic terms is motivated by the principle of causality, yet—in the final analysis—theism is forced to reject this principle. By positing an absolute beginning to the chain of causation, theists are implying that there can be something, at least one cause, which is itself outside the law of causality. This beginning, which is effectively the first cause, will itself be uncaused. This first cause will have to be an eternal and absolute principle. If so, how can one account for its capacity to produce things and events that are transient? Dharmakirti argues that no causal efficacy can be accorded to such a permanent principle. In essence, he is saying that the postulation of a first cause will have to be an arbitrary metaphysical hypothesis. It cannot be proven.

Asanga, writing in the fourth century, understood the origins of the universe in terms of the theory of dependent origination. This theory states that all things arise and come to an end in dependence upon causes and conditions. Asanga identifies three key conditions governing dependent origination. First is the condition of the *absence of a preceding intelligence*. Asanga rejects the possibility of the universe being the creation of a preceding intelligence, arguing that if one posits such an intelligence, it will have to totally transcend cause and effect. An absolute being that is eternal, transcendent, and beyond the domain of the law of causality would have no ability to interact with cause and effect, and therefore could neither start something nor stop it. Second is the condition of *impermanence*, which determines that the very causes and conditions that give rise to the world of dependent origination are themselves impermanent and subject to change. Third is the condition of *potentiality*. This principle refers to the fact that something cannot be produced from just anything. Rather, for a particular set of causes and conditions to give rise to a particular set of effects or consequences, there must be some kind of natural relationship between them. Asanga asserts that the origination of the universe must be understood in terms of the principle of an infinite chain of causation with no transcendence or preceding intelligence.

Buddhism and science share a fundamental reluctance to postulate a transcendent being as the origin of all things. This is hardly surprising given that both these investigative traditions are essentially nontheistic in their philosophical orientations. However, if on the one hand, the big bang is taken to be the absolute beginning, which implies that the universe has an absolute moment of origin, unless one refuses to speculate beyond this cosmic explosion, cosmologists must accept willy-nilly some kind of transcendent principle as the cause of the universe. This may not be the same God that the theists postulate; nonetheless, in its primary role as the creator of the universe, this transcendent principle will be a kind of godhead.

On the other hand, if (as some scientists have suggested) the big bang is less a starting point than a point of thermodynamic instability, there is room for a more nuanced and complex understanding of this cosmic event. I am told that many scientists feel the jury is still out as to whether the big bang is the absolute beginning of everything. The only conclusive empirical evidence so far, I am told, is that our own cosmic environment seems to have evolved from an intensely hot, dense state. Until more convincing evidence can be found for the various aspects of the big bang theory, and the key insights of quantum physics and the theory of relativity are fully integrated, many of the cosmological questions raised here will remain in the realm of metaphysics, not empirical science.

According to Buddhist cosmology, the world is constructed of the five elements: the supportive element of space, and the four basic elements of earth, water, fire, and air. Space enables the existence and functioning of all the other elements. The Kalachakra system presents space not as a total nothingness, but as a medium of "empty particles" or "space particles," which are thought of as extremely subtle "material" particles. This space element is the basis for the evolution and dissolution of the four elements, which are generated from it and absorbed back into it. The process of dissolution occurs in this order: earth, water, fire, and air. The process of generation occurs in this order: air, fire, water, and earth.

Asanga asserts that these basic elements, which he describes as the "four great elements," should not be understood in terms of materiality in the strict sense. He draws a distinction between the "four great elements," which are more like potentialities, and the four elements that are the constituents of aggregated matter. Perhaps the four elements within a material object may be better understood as solidity (earth), liquidity (water), heat (fire), and kinetic energy (air). The four elements are generated from the subtle level to the gross, from the underlying cause of the empty particles, and they dissolve from the gross level to the subtle and back into the empty particles of space. Space, with its empty particles, is the basis for the whole process. The term *particle* is perhaps not appropriate when referring to these phenomena, since it implies already formed material realities. Unfortunately, there is little description in the texts to help define these space particles further.

Buddhist cosmology establishes the cycle of the universe in the following way: first there is a period of formation, next a period when the universe endures, then a period when it is destroyed, followed by a period of void before the formation of a new universe. During the fourth period, that of emptiness, the space particles subsist, and it is from these particles that all the matter within a new universe is formed. It is in these space particles that we find the fundamental cause of the entire physical world. If we wish to describe the formation of the universe and the physical bodies of beings, we need to analyze the way the different elements constituting that universe were able to take shape from these space particles.

It is on the basis of the specific potential of those particles that the structure of the universe and everything in it—planets, stars, sentient beings, such as humans and animals—have come about. If we go back to the ultimate cause of the material objects of the world, we arrive finally at the space particles. They precede the big bang (which is to say any new beginning) and are indeed the residue of the preceding universe that has disintegrated. I am told that some cosmologists favor the idea that our universe arose as a fluctuation from what is termed the *quantum vacuum*. To me, this idea echoes the Kalachakra theory of space particles.

From the point of view of modern cosmology, understanding the origin of the universe during the first few seconds poses an almost insurmountable challenge. Part of the problem lies in the fact that the four known forces of nature—gravitation and electromagnetism, and the weak and strong nuclear forces—are not functioning at this point. They come into play later, when the density and temperature of the initial stage have significantly decreased so that the elementary particles of matter, such as hydrogen and helium, begin to form. The exact beginning of the big bang is what is called a "singularity." Here, all mathematical equations and laws of physics break down. Quantities that are normally measurable, such as density and temperature, become undefined at such a moment.

Since scientific study of cosmological origin requires the application of mathematical equations and the assumption of the validity of the laws of physics, it would seem that, if these equations and laws break down, we must ask ourselves whether we can ever have a complete understanding of the initial few seconds of the big bang. My scientist friends have told me that some of the best minds are engaged in exploring the story of the first stages of the formation of our universe. I am told that some believe the solution to what currently appears as a set of insurmountable problems must lie in finding a grand unified theory, which will help integrate all the known laws of physics. Perhaps it can bring together the two paradigms of modern physics that seem to contradict each other—relativity and quantum mechanics. I am told that the axiomatic assumptions of these two theories have so far proven impossible to reconcile. The theory of relativity suggests that the accurate calculation of the precise condition of the cosmos at any given time is possible if one has sufficient information. Quantum mechanics, by contrast, asserts that the world of microscopic particles can be understood only in probabilistic terms, because at a fundamental level the world consists of chunks or quanta of matter (hence the name *quantum physics*), which are subject to the uncertainty principle. Theories with exotic names like superstring theory or the M theory are being proposed as candidates for the grand unified theory.

There is a further challenge to the very enterprise of obtaining full knowledge of the original unfolding of our universe. At the fundamental level, quantum mechanics tells us that it is impossible to predict accurately how a particle might behave in a given situation. One can, therefore, make predictions about the behavior of particles only on the basis of probability. If this is so, no matter how powerful one's mathematical formulas might be, since our knowledge of the initial conditions of a given phenomenon or an event will always be incomplete, we cannot fully understand how the rest of the story unfolds. At best, we can make approximate conjectures, but we can never arrive at a complete description even of a single atom, let alone the entire universe.

In the Buddhist world, there is an acknowledgment of the practical impossibility of gaining total knowledge of the origin of the universe. A Mahayana text entitled *The Flower Ornament Scripture* contains a lengthy discussion of infinite world systems and the limits of human knowledge. A section called "The Incalculable" provides a string of calculations of extremely high numbers, culminating in terms such as "the incalculable," "the measureless," "the boundless," and "the incomparable." The highest number is the "square untold," which is said to be the function of the "unspeakable" multiplied by itself! A friend told me that this number can be written as 10^{59}. The *Flower Ornament* goes on to apply these mind-boggling numbers to the universe systems; it suggests that if "un-told" worlds are reduced to atoms and each atom contains "untold" worlds, still the numbers of world systems will not be exhausted.

Similarly, in beautiful poetic verses, the text compares the in-tricate and profoundly interconnected reality of the world to an in-finite net of gems called "Indra's jeweled net," which reaches out to infinite space. At each knot on the net is a crystal gem, which is connected to all the other gems and reflects in itself all the others. On such a net, no jewel is in the center or at the edge. Each and every jewel is at the center in that it reflects all the other jewels on the net. At the same time, it is at the edge in that it is itself reflected in all the other jewels. Given the profound interconnectedness of everything in the universe, it is not possible to have total knowl-edge of even a single atom unless one is omniscient. To know even one atom fully would imply knowledge of its relations to all other phenomena in the infinite universe.

The Kalachakra texts claim that, prior to its formation, any par-ticular universe remains in the state of emptiness, where all its ma-terial elements exist in the form of potentiality as "space particles." At a certain point, when the karmic propensities of the sentient be-ings who are likely to evolve in this particular universe ripen, the "air particles" begin to aggregate with each other, creating a cosmic wind. Next the "fire particles" aggregate in the same way, creating powerful "thermal" charges that travel through the air. Following this, the "water particles" aggregate to form torrential "rain" accom-panied by lightning. Finally, the "earth particles" aggregate and, combined with the other elements, begin to assume the form of so-lidity. The fifth element, "space," is thought to pervade all other el-ements as an immanent force and therefore does not possess a distinct existence. Over a long temporal process, these five elements expand to form the physical universe as we come to know and experience it.

So far we have been speaking of the origin of the universe as though it consisted only of a mix of inanimate matter and energy— the birth of galaxies, black holes, stars, planets, and the wildness of subatomic particles. From the Buddhist perspective, however, there is the critical issue of the role of consciousness. For example, inherent in both the Kalachakra and Abhidharma cosmologies is the idea that the formation of a particular universe system is intimately connected with the karmic propensities of sentient beings. In contemporary language, these Buddhist cosmologies can be seen as proposing that our planet evolved in such a way that it could support the evolution of sentient beings in the forms of the myriads of species that exist today on the earth.

By invoking karma here, I am not suggesting that according to Buddhism everything is a function of karma. We must distinguish between the operation of the natural law of causality, by which once a certain set of conditions are put in motion they will have a certain set of effects, and the law of karma, by which an intentional act will reap certain fruits. So, for example, if a campfire is left in a forest and catches onto some dry twigs, leading to a forest fire, the fact that once the trees are aflame they burn, becoming charcoal and smoke, is simply the operation of the natural law of causality, given the nature of fire and the materials that are burning. There is no karma involved in this sequence of events. But a sentient being choosing to light a campfire and forgetting to put it out—which began the chain of events—here karmic causation is involved.

My own view is that the entire process of the unfolding of a universe system is a matter of the natural law of causality. I envision karma coming into the picture at two points. When the universe has evolved to a stage where it can support the life of sentient beings, its fate becomes entangled with the karma of the beings who will inhabit it. More difficult perhaps is the first intervention of karma, which is effectively the maturation of the karmic potential of the sentient beings who will occupy that universe, which sets in motion its coming into being.

The ability to discern exactly where karma intersects with the natural law of causation is traditionally said to lie only within the Buddha's omniscient mind. The problem is how to reconcile two strands of explanation—first, that any universe system and the beings within it arise from karma, and second, that there is a natural process of cause and effect, which simply unfolds. The early Buddhist texts suggest that matter on the one hand and consciousness on the other relate according to their own process of cause and effect, which gives rise to new sets of functions and properties in both cases. On the basis of understanding their nature, causal relations, and functions, one can then derive inferences—for both matter and consciousness—that give rise to knowledge. These stages were codified as "four principles"—the principle of nature, the principle of dependence, the principle of function, and the principle of evidence.

The question then is, Are these four principles (which effectively constitute the laws of nature according to Buddhist philosophy) themselves independent of karma, or is even their existence tied to the karma of the beings that inhabit the universe in which they operate? This issue is analogous to the questions raised in relation to the status of the laws of physics. Can there be a completely different set of laws of physics in a different universe, or do the laws of physics as we understand them hold true in all possible universes? If the answer is that a different set of laws can operate in a different universe system, this would suggest (from a Buddhist perspective) that even the laws of physics are entangled with the karma of the sentient beings that will arise in that universe.

How do Buddhist cosmological theories envision the unfolding of the relationship between the karmic propensities of sentient beings and the evolution of a physical universe? What is the mechanism by which karma connects to the evolution of a physical system? On the whole, the Buddhist Abhidharma texts do not have much to say on these questions, apart from the general point that the environment where a sentient being exists is an "environmental effect" of the being's collective karma shared with myriad other beings. However, in the Kalachakra texts, close correlations are drawn between the cosmos and the bodies of the sentient beings living in it, between the natural elements in the external physical universe and the elements within the bodies of sentient beings, and between the phases in the passage of the celestial bodies and the changes within the bodies of the sentient beings. The Kalachakra presents a detailed picture of these correlations and their manifestations in the experience of a sentient creature. For example, the texts speak of how solar and lunar eclipses may affect the body of a sentient being through changing patterns in breathing. It would be interesting to subject some of these claims, which are empirical, to scientific investigation.

Even with all these profound scientific theories of the origin of the universe, I am left with questions, serious ones: What existed before the big bang? Where did the big bang come from? What caused it? Why has our planet evolved to support life? What is the relationship between the cosmos and the beings that have evolved within it? Scientists may dismiss these questions as nonsensical, or they may acknowledge their importance but deny that they belong to the domain of scientific inquiry. However, both these approaches will have the consequence of acknowledging definite limits to our scientific knowledge of the origin of our cosmos. I am not subject to the professional or ideological constraints of a radically materialistic worldview. And in Buddhism the universe is seen as infinite and beginningless, so I am quite happy to venture beyond the big bang and speculate about possible states of affairs before it.

5

·

Evolution, Karma, and the World of Sentience

The question What is life?, regardless of how it may be framed, poses a challenge to any intellectual attempt to develop a coherent worldview. Like modern science, Buddhism holds the basic premise that, at the most fundamental level, there is no qualitative difference between the material basis of the body of a sentient being, such as a human, and that of, say, a piece of rock. Just as a rock is constituted by an aggregation of material particles, the human body is composed of similar material particles. Indeed, the entire cosmos and all the matter in it are made from the same stuff, which is endlessly recycled—according to science, the atoms in our body once belonged to stars far away in time and space.

The question then is, What makes a human body so different from a rock that it can support life and consciousness? The modern biological response to this question turns on the notion of the emergence of higher levels of properties corresponding to higher levels of complexity in the aggregation of the material constituents. In other words, modern biology tells the story through an increasingly complex aggregation of atoms into molecular and genetic structures; the complex organism of life emerges simply on the basis of material elements.

Darwinian evolution is the conceptual underpinning of modern biology. The theory of evolution, and in particular the notion of natural selection, provides the big picture of the origin of diverse life-forms. As I understand it, the theories of evolution and natural selection are attempts to account for the miraculous variety of living things. The spectacular richness of life and the huge differences among the many species are explained by the scientific idea that new forms are created by the alteration of present forms, with the added idea that those features best suited to a given environment will be passed on to subsequent generations, while those features not essential to survival die out.

These theories describe, I have been told, what Darwin himself called a "descent" into the multiplicity and complexity of all forms of life from an original simplicity. Since all living beings belong to evolutionary lineages stretching back to a common ancestor, the theory stresses the original interconnectedness of living beings in the world.

I heard about the theory of evolution when I made my first trip to India in 1956, and it was there that I was exposed to some of the theoretical aspects of modern biology. But it was only much later that I was able to speak to a real scientist at length about Darwin's evolutionary theory. Ironically, the first person to help me understand the theory more fully was not a scientist but a scholar of religion. Huston Smith came to see me in Dharamsala in the 1960s. We spoke about the world religions, the need for greater pluralism among their followers, and the role of spirituality in an increasingly materialistic world, as well as some more esoteric reflections on possible areas of convergence between Buddhist and Christian mysticism. However, the topic that struck me most was modern biology, especially our discussion of DNA and the fact that so many secrets of life appear to lie in the mystery of this beautiful biological string. When I count my teachers of science, I include Huston Smith among them, although I am not sure whether he would himself approve of this.

The exponential rate of progress in biology, especially the revolution in genetic science, has radically deepened our understanding of the role of DNA in unlocking the mysteries of life. My own understanding of modern biology owes much to the counsel of great teachers like the late Robert Livingston from the University of California at San Diego. He was a very patient teacher who peered intensely through his glasses as he explained a point and a passionately caring man with a deep commitment to world nuclear disarmament. Among his gifts to me was a plastic model of the brain with detachable labeled components, which today sits on my desk in Dharamsala, and a handwritten synopsis of the key points of neurobiology.

The Darwinian theory is an explanatory framework that accounts for the wealth of flora and fauna, the richness of what Buddhists call sentient beings and plants that effectively constitute the biological world available to us. So far the theory has avoided disproof and has offered the most coherent scientific understanding of the evolution of the diversity of life on earth. The theory applies as much to the molecular level—that is, to the adaptation and selection of individual genes—as to the macrocosmic level of large organisms. Despite its remarkable adaptability to all levels on which we might say life flourishes, Darwin's theory does not explicitly address the conceptual question of what life is. This said, there are a number of key characteristics that biology understands to be essential for life, such as organisms being self-sustaining systems and naturally possessing some mechanisms for reproduction. In addition, the key definitions of life include the ability to develop away from chaos and toward order, which is called "negative entropy."

The Buddhist Abhidharma tradition, by contrast, defines *sok*, the Tibetan equivalent for the English term *life,* as that which supports "heat" and "consciousness." To some extent the differences are semantic, since what Buddhist thinkers mean by *life* and living relates entirely to sentient beings and not to plants, while modern biology has a much broader conception of life, taking it all the way down to the cellular level. The Abhidharma definition does not correspond to the biological account mostly because the underlying motive of Buddhist theory is to answer ethical questions that can be considered only in relation to higher forms of life.

Central to Darwin's theory of evolution, as I understand it, is natural selection. But what does this mean? The biological model represents natural selection as random genetic mutation and subsequent competition between organisms leading to the "survival of the fittest" or, more correctly, the differential reproductive success of some organisms versus others. Every trait in an organism is subjected to the constraints of the environment. Those organisms which thrive best within these constraints and in competition with others, and which have the most offspring, are deemed better adapted and thus better equipped to survive. As the most suitable features are continuously selected in a given environment from among the variations produced by random mutations, the species of living beings transform.

Natural selection can be seen as explaining which kinds of flies or monkeys can best survive in their chosen environments, and how beings like modern humans evolved from apelike ancestors. Despite their obvious differences, humans and chimpanzees share 98 percent of their DNA; a difference of only 2 percent accounts for the distinction between the two species (the difference between humans and gorillas is 3 percent). Likewise, on the genetic level, natural selection seems to explain how mutations in genes, which are random but arise naturally, can be selected out and hence create new varieties within living beings. Genetic mutation is also thought to be the engine for evolution at the molecular level. And natural selection is seen as the mechanism that favors the development of neural groups (transmitters, receptors, and so forth), which give rise to the individuality and variability of each brain and, on the level of the species, to the special qualities of human consciousness, for example.

Even in relation to the origins of life, natural selection is taken as the key to a process whereby particular molecules capable of replicating themselves arose (perhaps initially by chance) in an organic primeval "soup" or possibly as self-replicating inorganic crystals. Indeed, I gather from the Stanford physicist Stephen Chu that his team is currently developing models for understanding life in terms of the laws of physics. According to the current story of the origins of organic life, shortly after the earth itself came into being, molecules of RNA (ribonucleic acid), themselves highly unstable, came into being and self-replicated without assistance. By natural selection, tougher and more durable molecules—molecules of DNA (deoxyribonucleic acid, the fundamental repository of genetic information)—emerged from the RNA. Life came into being in the form of a more sophisticated creature that stored the genetic recipe for its makeup in DNA and made its form from protein. RNA became the link between DNA and protein, since it reads the information stored in DNA and guides the production of proteins.

The first organism composed of DNA, RNA, and protein is known as Luca, the last universal common ancestor, which may have been something like a bacterium living deep in the earth or in warm water. Again, through self-replication and by natural selection, Luca gradually evolved into all beings. I always smile when I hear this name, as Luca is the name of my longtime Italian translator.

This model presupposes a pattern of small and gradual changes that lead to innumerable variations in living beings. These variations are what are screened by natural selection. There are various alternatives to this picture—for instance, the possibility of large and sudden changes, and hence a view of evolution developing through leaps in which the transformations of organisms are not piecemeal but dramatic. Likewise, there is a debate about whether natural selection is the sole mechanism of change or whether other factors are involved as well.

The explosion in genetic science of very recent times has given incomparably greater sophistication and specificity to our understanding of evolution at the molecular and genetic level. With immaculate timing, just before the fiftieth anniversary of the discovery of the structure of DNA by James Watson and Francis Crick in 1953, the sequencing of the human genome was completed. This colossal feat carried the promise of untold technological and medical potential.

I heard about the sequencing of the genome in an unusual way. On the day the American president Bill Clinton and the British prime minister Tony Blair jointly made the announcement, I was in the United States and scheduled to appear on the *Larry King Live* show. Since I hear the news only early in the morning or at the end of the day, I had missed the announcement that afternoon. So when Larry King asked me what I thought of it, I hadn't the faintest idea what he was talking about. Somehow, I couldn't connect the announcement of a scientific breakthrough of this magnitude with two politicians making press statements. The fact that my interview was conducted through a satellite linkup did not make the conversation any easier. So it was Larry King on live television who broke the news to me.

The broader implications of this amazing scientific feat are becoming increasingly felt: I have had the opportunity to talk with scientists working in the field, especially the geneticist Eric Lander at MIT. He showed me around his laboratory at the Broad Institute of MIT and Harvard, where many of the powerful machines used for sequencing the human genome are at work, and demonstrated some of the stages involved in sequencing a genome.

At one of the Mind and Life conferences, Eric explained the human genome by comparing it to the *kangyur,* the collection of scriptures attributed to the Buddha and translated into Tibetan, which has just over a hundred volumes of about three hundred folios each. The massive book of the genome has twenty-three chapters, the twenty-three human chromosomes, and each set of the genome (one set from each parent) contains between thirty thousand and eighty thousand genes. Each of these chapters is written on a long chain of DNA in three-letter words, which are composed of the four letters A, C, G, and T—adenine, cytosine, guanine, and thymine—sequenced in all possible combinations.

Imagine, Eric suggested, that during the millions of years of copying this book, every now and then some small errors creep in, just as—in the hundreds of years of copying it by hand—small scribal errors, misspellings, and substitutions of words enter the text of the *kangyur*. These errors may be perpetuated in subsequent copying, which then introduces new copying variations, and so on. Some of these spelling variations may not have a radical impact on the reading of the text; however, sometimes there occurs a crucial spelling error that can have far-reaching consequences. In the analogy to a canonical text, although the change may be a single spelling error, if this is, say, from a positive to a negative word, there may be a radical effect on the meaning of a sentence or the reading of the whole text. It is these random variations in spellings, I am told, which are the mutations that occur naturally in the evolutionary process.

According to some biologists I have spoken with, there is growing consensus that the occurrence of genetic mutations, regardless of how natural they may be, remains entirely random. However, once such mutations take place, the principle of natural selection ensures that, on the whole, those mutations or changes that promote the best chance for survival get selected. As the American biologist Ursula Goodenough aptly put it at a Mind and Life conference in 2002, "Mutation is utterly random, but selection is extremely choosy!" From a philosophical point of view, the idea that these mutations, which have such far-reaching implications, take place naturally is unproblematic, but that they are purely random strikes me as unsatisfying. It leaves open the question of whether this randomness is best understood as an objective feature of reality or better understood as indicating some kind of hidden causality.

By contrast with science, in Buddhism there is no substantive philosophical discussion on how living organisms emerge from inanimate matter. In fact, there does not appear even to be an acknowledgment that this is a serious philosophical issue. At best there is an implicit assumption that the emergence of living organisms from inanimate matter is simply a consequence of cause and effect over time, given a set of initial conditions and the laws of nature that govern all realms of existence. However, in Buddhism there is a greater appreciation of the challenge of accounting for the emergence of sentient beings from what is essentially a non-sentient basis.

This difference of concern suggests an interesting contrast between Buddhism and modern science, which may have partly to do with the complex historical, social, and cultural differences that underlie the development of these two investigative traditions. For modern science, at least from a philosophical point of view, the critical divide seems to be between inanimate matter and the origin of living organisms, while for Buddhism the critical divide is between non-sentient matter and the emergence of sentient beings.

We may even ask why there is this fundamental difference between the two traditions. One possible reason modern science perceives the critical divide to lie between inanimate matter and living organisms may have to do with the basic methodology of science. By this I am referring to reductionism, not so much as a metaphysical standpoint but more as a methodological approach. The basic approach in science is to explain phenomena in terms of their simpler constitutive elements. How can something like life emerge from non-life? At one of the Mind and Life conferences in Dharamsala, the Italian biologist Luigi Luisi, based in Zurich, told me of his team's research into the possibility of creating life in a laboratory. For if the current scientific theory of the origination of life from the complex configuration of inanimate matter is correct, there is nothing to prevent us from creating life in a lab once all the conditions are met.

Buddhism draws the critical division differently—i.e., between sentience and non-sentience—because it is primarily interested in the alleviation of suffering and the quest for happiness. In Buddhism, the evolution of the cosmos and the emergence of the sentient beings within it—indeed, effectively everything within the purview of the physical and life sciences—belong within the domain of the first of the Four Noble Truths, which the Buddha taught in his initial sermon. The Four Noble Truths state that within the realm of impermanent phenomena there is suffering, suffering has an origin, the cessation of suffering is possible, and there is a path to the cessation of suffering. As I see it, science falls within the scope of the first truth in that it examines the material bases of suffering, for it covers the entire spectrum of the physical environment—"the container"—as well as the sentient beings—"the contained." It is in the mental realm—the realm of psychology, consciousness, the afflictions, and karma—that we find the second of the truths, the origin of suffering. The third and fourth truths, cessation and the path, are effectively outside the domain of scientific analysis in that they pertain primarily to what might be called philosophy and religion.

This fundamental difference between Buddhism and science—whether the line is drawn between sentience and non-sentience or between living organisms and inanimate matter—has significant ramifications, among them a difference in how the two investigative traditions may regard consciousness. For biology, consciousness is a secondary issue, since it is a characteristic of a subset of living organisms rather than of all of life. In Buddhism, since the definition of "living" refers to sentient beings, consciousness is the primary characteristic of "life."

One implicit assumption I have sometimes found in Western thought is that, in the story of evolution, human beings enjoy a unique existential status. This uniqueness is often understood in terms of some kind of "soul" or "self-consciousness," which only humans are thought to possess. Many people appear to assume implicitly three incremental stages in the development of life: inanimate matter, living organisms, and human beings. Behind this view may lie an idea that human beings occupy a distinctly different category from animals and plants. Strictly speaking, this is not a scientific concept.

In contrast, if one examines the history of Buddhist philosophical thinking, there is an understanding that animals are closer to humans (in that both are sentient beings) than they are to plants. This understanding is based on the notion that, insofar as their sentience is concerned, there is no difference between humans and animals. Just as we humans wish to escape suffering and to seek happiness, so do animals. Similarly, just as we humans have the capacity to experience pain and pleasure, so do animals. Philosophically speaking, from the Buddhist point of view, both human beings and animals possess what in Tibetan is called *shepa*, which can be roughly translated as "consciousness," albeit to different degrees of complexity. In Buddhism, there is no recognition of the presence of something like the "soul" that is unique to humans. From the perspective of consciousness, the difference between humans and animals is a matter of degree and not of kind.

In the earliest Buddhist scriptures there is an allusion to a story of human evolution, which is recounted in many of the subsequent Abhidharma texts. The story unfolds in the following manner. The Buddhist cosmos consists of three realms of existence—the desire realm, the form realm, and the formless realm—the last being progressively subtler states of existence. The desire realm is characterized by the experience of sensual desires and pain; this is the realm that we humans and animals inhabit. In contrast, the form realm is free from any manifest experience of pain and is permeated principally by an experience of bliss. Beings in this realm possess bodies composed of light. Finally, the formless realm utterly transcends all physical sensation. Existence in this realm is permeated by an abiding state of perfect equanimity, and the beings in this realm are entirely free from material embodiment. They exist only on an immaterial mental plane. Beings in the higher states of the desire realm and those in both the form and formless realms are described as celestial beings. It should be noted that these realms also fall within the first noble truth. They are not permanent, heavenly states to which we should aspire. They come with their own suffering of impermanence.

The evolution of human life on earth is understood in terms of the "descent" of some of these celestial beings, who have exhausted their positive karma, which provided them with the cause and conditions to remain in these higher realms. There was no original sin that precipitated the fall; it's simply the nature of impermanent existence, cause and effect, that causes a being to change states, "to die." When these beings first experienced their "fall" and were born on earth, they still possessed vestiges of their previous glories. These humans of the first era were thought to have godlike qualities. They are said to have come into being through "spontaneous birth," they had attractive physiques, their bodies had halos, they had certain supernormal powers, like flying, and they subsisted on the nourishment of inner contemplation. They were also thought to be free from many of the features that serve as the basis for discriminating identity, such as gender, race, and caste.

Over time, it is said, humans began to lose these qualities. As they took nourishment from material food, their bodies assumed coarser corporeality and thus gave rise to a great diversity of physical appearances. This diversity in turn led to feelings of discrimination, especially animosity toward those who appeared different and attachment toward those who were similar, resulting in the emergence of the whole host of gross negative emotions. Furthermore, the dependence on material food led to the need for the disposal of waste from the body, and—I am not quite sure of how the reasoning works here—this need led to the emergence of the male and female sex organs on the human body. The story continues with a detailed account of the genesis of the entire range of negative human actions, such as killing, stealing, and sexual misconduct.

Central to this account of human evolution is the Abhidharma theory of the four types of birth. In this view, sentient beings can come into being as (1) womb-born, such as we humans; (2) egg-born, such as the birds and many reptiles; (3) heat and moisture-born, such as the numerous types of insects; and (4) spontaneously born, such as the celestial beings in the form and formless realms. As to the question of diversity of life, Chandrakirti expressed a common Buddhist viewpoint when he wrote, "It is from the mind that the world of sentience arises. So too from the mind the diverse habitats of beings arise."

In the earliest scriptures attributed to the Buddha, we find similar statements on how, ultimately, mind is the creator of the entire universe. There have been Buddhist schools that took such statements literally and adopted a radical form of idealism whereby the reality of the external material world is rejected. But on the whole, most Buddhist thinkers have tended to interpret such statements as meaning that we must understand the origination of the world, at least the world of sentient creatures, through the activity of karma.

The theory of karma is of signal importance in Buddhist thought but is easily misrepresented. Literally, *karma* means "action" and refers to the intentional acts of sentient beings. Such acts may be physical, verbal, or mental—even just thoughts or feelings—all of which have impacts upon the psyche of an individual, no matter how minute. Intentions result in acts, which result in effects that condition the mind toward certain traits and propensities, all of which may give rise to further intentions and actions. The entire process is seen as an endless self-perpetuating dynamic. The chain reaction of interlocking causes and effects operates not only in individuals but also for groups and societies, not just in one lifetime but across many lifetimes.

When we use the term *karma*, we may refer both to specific and individual acts and to the whole principle of such causation. In Buddhism, this karmic causality is seen as a fundamental natural process and not as any kind of divine mechanism or working out of a preordained design. Apart from the karma of individual sentient beings, whether it is collective or personal, it is entirely erroneous to think of karma as some transcendental unitary entity that acts like a god in a theistic system or a determinist law by which a person's life is fated. From the scientific view, the theory of karma may be a metaphysical assumption—but it is no more so than the assumption that all of life is material and originated out of pure chance.

As to what might be the mechanism through which karma plays a causal role in the evolution of sentience, I find helpful some of the explanations given in the Vajrayana traditions, often referred to by modern writers as esoteric Buddhism. According to the Guhyasamaja tantra, a principal tradition within Vajrayana Buddhism, at the most fundamental level, no absolute division can be made between mind and matter. Matter in its subtlest form is *prana*, a vital energy which is inseparable from consciousness. These two are different aspects of an indivisible reality. *Prana* is the aspect of mobility, dynamism, and cohesion, while consciousness is the aspect of cognition and the capacity for reflective thinking. So according to the Guhyasamaja tantra, when a world system comes into being, we are witnessing the play of this energy and consciousness reality.

Because of this indivisibility of consciousness and energy, there is a profoundly intimate correlation between the elements within our bodies and the natural elements in the outside world. This subtle connection can be discerned by individuals who have gained a certain level of spiritual realization or who have a naturally higher level of perception. For example, the fifteenth-century Tibetan thinker Taktsang Lotsawa conducted an experiment upon himself and found a total concordance between his personal experience of changes that naturally occur in one's breathing pattern and those described in the Kalachakra tantra during a celestial event like solar or lunar eclipse. In fact, according to Buddhist Vajrayana thought, there is an understanding that our bodies represent microcosmic images of the greater, macrocosmic world. Because of this perspective, the Kalachakra tantra pays tremendous attention to study of the celestial bodies and their movements; in fact, there is an elaborate system of astronomy in these texts.

Just as I never found the Abhidharma cosmology convincing, I have never really been persuaded by the Abhidharma account of human evolution as progressive "degeneration." One of Tibet's own myths of creation tells how the Tibetan people evolved from the mating of a monkey and a fierce ogress, and of course I'm not convinced by that either!

On the whole, I think the Darwinian theory of evolution, at least with the additional insights of modern genetics, gives us a fairly coherent account of the evolution of human life on earth. At the same time, I believe that karma can have a central role in understanding the origination of what Buddhism calls "sentience," through the media of energy and consciousness.

Despite the success of the Darwinian narrative, I do not believe that all the elements of the story are in place. To begin with, although Darwin's theory gives a coherent account of the development of life on this planet and the various principles underlying it, such as natural selection, I am not persuaded that it answers the fundamental question of the origin of life. Darwin himself, I gather, did not see this as an issue. Furthermore, there appears to be a certain circularity in the notion of "survival of the fittest." The theory of natural selection maintains that, of the random mutations that occur in the genes of a given species, those that promote the greatest chance of survival are most likely to succeed. However, the only way this hypothesis can be verified is to observe the characteristics of those mutations that have survived. So in a sense, we are stating simply this: "Because these genetic mutations have survived, they are the ones that had the greatest chance of survival."

From the Buddhist perspective, the idea of these mutations being purely random events is deeply unsatisfying for a theory that purports to explain the origin of life. Karl Popper once commented that, to his mind, Darwin's theory of evolution does not and cannot explain the origin of life on earth. For him, the theory of evolution is not a testable scientific theory but rather a metaphysical theory that is highly beneficial for guiding further scientific research. Moreover, the Darwinian theory, while acknowledging the critical distinction between inanimate matter and living organisms, fails to acknowledge adequate qualitative distinctions between living organisms, such as trees and plants on the one hand and sentient creatures on the other.

One empirical problem in Darwinism's focus on the competitive survival of individuals, which is defined in terms of an organism's struggle for individual reproductive success, has consistently been how to explain altruism, whether in the sense of collaborative behavior, such as food sharing or conflict resolution among animals like chimpanzees or acts of self-sacrifice. There are many examples, not only among human beings but among other species as well, of individuals who put themselves in danger to save others. For instance, a honeybee will sting to protect its hive from intruders, even though the act of stinging causes it to die; or the Arabian babbler, a type of bird, will risk its own safety to warn the rest of the flock of an attack.

Post-Darwinian theory has attempted to explain such phenomena by arguing that there are circumstances in which altruistic behavior, including self-sacrifice, enhances an individual's chances of passing on its genes to future generations. However, I do not think this argument applies to instances where, I am told, altruism can be observed across species. For example, one might think of the host birds who wean and rear baby cuckoos left in their nests, although some have explained this exclusively in terms of the selfish benefit obtained by the cuckoos. Moreover, given that this kind of altruism does not always appear to be voluntary—some organisms seem programmed to act in a self-sacrificial way—modern biology would basically see altruism as instinctive and dictated by genes. The problem becomes all the more complex if we bring in the question of human emotion, especially the numerous instances of altruism in human society.

Some more dogmatic Darwinians have suggested that natural selection and survival of the fittest are best understood at the level of individual genes. Here we see the reduction of the strong metaphysical belief in the principle of self-interest to imply that somehow individual genes behave in a selfish way. I do not know how many of today's scientists hold such radical views. As it stands, the current biological model does not allow for the possibility of real altruism.

At one of the Mind and Life conferences in Dharamsala, the Harvard historian of science Anne Harrington made a memorable presentation on how, and to some extent why, scientific investigation of human behavior has so far failed to develop any systematic understanding of the powerful emotion of compassion. At least in modern psychology, compared with the tremendous amount of attention paid to negative emotions, such as aggression, anger, and fear, relatively little examination has been made of more positive emotions, such as compassion and altruism. This emphasis may have arisen because the principal motive in modern psychology has been to understand human pathologies for therapeutic purposes. However, I do feel that it is unacceptable to reject altruism

on the ground that selfless acts do not fit within the current biological understanding of life or are simply redefinable as expressions of the self-interest of the species. This stance is contrary to
the very spirit of scientific inquiry. As I understand it, the scientific
approach is not to modify the empirical facts to fit one's theory;
rather the theory must be adapted to fit the results of empirical inquiry. Otherwise it would be like trying to reshape one's feet to fit
the shoes.

I feel that this inability or unwillingness fully to engage the
question of altruism is perhaps the most important drawback of
Darwinian evolutionary theory, at least in its popular version. In
the natural world, which is purported to be the source of the theory of evolution, just as we observe competition between and
within species for survival, we observe profound levels of cooperation (not necessarily in the conscious sense of the term). Likewise,
just as we observe acts of aggression in animals and humans, we
observe acts of altruism and compassion. Why does modern biology accept only competition to be the fundamental operating principle and only aggression to be the fundamental trait of living
beings? Why does it reject cooperation as an operating principle,
and why does it not see altruism and compassion as possible traits
for the development of living beings as well?

To what extent we should ground the entirety of our conception of human nature and existence in science depends, I suppose,
on what conception of science we hold. For me, this is not a scientific question; rather it is a matter of philosophical persuasion. A
radical materialist might wish to support the thesis that evolutionary theory accounts for all aspects of human life, including moral-
ity and religious experience, while others might perceive science as
occupying a more limited scope in understanding the nature of human existence. Science may never be able to tell us the full story of

dhist philosophers also speak of *sem* (*citta* in Sanskrit), "mind" in English; *namshe* (*vijnana* in Sanskrit), "consciousness"; and *yi* (*manas* in Sanskrit), "mentality" or "mental states."

The Tibetan word *namshe*, or its Sanskrit equivalent, *vijnana*, which is often translated as "consciousness," has a broader range of application than the English term in that it covers not only the whole range of conscious experiences but also those forces that might be recognized as part of the so-called unconscious according to modern psychological and psychoanalytic theories. Furthermore, the Tibetan word for "mind," which is *sem* (Sanskrit *citta*), covers not just the realm of thought but also that of emotion. We can speak of the phenomena of consciousness without excessive confusion, but we need to be heedful of the limitations of our respective linguistic terms.

The problem of describing the subjective experiences of consciousness is complex indeed. For we risk objectivizing what is essentially an internal set of experiences and excluding the necessary presence of the experiencer. We cannot remove ourselves from the equation. No scientific description of the neural mechanisms of color discrimination can make one understand what it feels like to perceive, say, the color red. We have a unique case of inquiry: the object of our study is mental, that which examines it is mental, and the very medium by which the study is undertaken is mental. The question is whether the problems posed by this situation for a scientific study of consciousness are insurmountable—are they so damaging as to throw serious doubt on the validity of the inquiry?

Although we tend to relate to the mental world as if it were homogenous—a somewhat monolithic entity called "the mind"—when we probe more deeply, we come to recognize that this approach is too simplistic. As we experience it, consciousness is made up of myriad highly varied and often intense mental states.

need an object—something to be conscious *of*? What is its relation to the unconscious—not only the unconscious electrochemical events of the brain that are correlated with mental processes but also more complex and perhaps problematic unconscious desires, memories, and expectations? Given the highly subjective nature of our experience of consciousness, is a scientific understanding—in the sense of an objective, third-person account—ever possible?

The question of consciousness has attracted a good deal of attention in the long history of Buddhist philosophical thinking. For Buddhism, given its primary interest in questions of ethics, spirituality, and overcoming suffering, understanding consciousness, which is thought to be a defining characteristic of sentience, is of great importance. According to the earliest scriptures, the Buddha saw consciousness as playing a key role in determining the course of human happiness and suffering. For example, the famous discourse of the Buddha known as the *Dhammapada* opens with the statement that mind is primary and pervades all things.

Before we proceed, it is important to be aware of the problems raised by our use of language in describing subjective experience. Despite the universality of the experience of consciousness, the languages in which we articulate our subjective experiences have their roots in disparate cultural, historical, and linguistic backgrounds. These diverse backgrounds represent different cognitive frameworks—conceptual mapping, linguistic practices, or philosophical and spiritual heritage. For example, in Western European languages, one speaks of "consciousness," "the mind," "mental phenomena," and "awareness." Similarly, in the context of Buddhist philosophy of mind, one speaks of *lo* (*buddhi* in Sanskrit), *shepa* (*jñana*), and *rigpa* (*vidya*)—all of which can be roughly translated as awareness or "intelligence" in the broadest sense of the term. Bud-

method—the objective perspective from the outside—has made strikingly little headway in this understanding.

There is, however, a growing recognition that the study of consciousness is becoming a most exciting area of scientific investigation. At the same time, there is a growing acknowledgment that modern science does not yet possess a fully developed methodology to investigate the phenomenon of consciousness. This is not to say that there have been no philosophical theories on the subject, or that there have been no efforts to "explain" consciousness in terms of material paradigms. At one extreme was the standpoint of behaviorism, which attempted to define consciousness in terms of the language of external behavior, thus reducing mental phenomena to verbal and bodily action. At the other extreme was what is known as Cartesian dualism, the idea that the world comprises two independent, substantially real things—matter, which is characterized by qualities such as extension, and mind, which is defined in terms of an immaterial substance, such as the "spirit." Between these two extremes all kinds of theories have been proposed, from functionalism (which attempts to define consciousness in terms of its functions) to neurophenomenology (which attempts to define consciousness in terms of neural correlates). Most of these theories understand consciousness by means of aspects of the material world.

But what about the direct observation of consciousness itself? What are its characteristics and how does it function? Does all of life (plants as well as animals) share in it? Does our conscious life exist only when we are aware of being conscious, so that in dreamless sleep, for example, consciousness may be said to be dormant, or even annihilated? Is consciousness composed of serial moments of mental fluctuation, or is it continuous but continually changing? Is consciousness a matter of degree? Does consciousness always

The joy of meeting someone you love, the sadness of losing a close friend, the richness of a vivid dream, the serenity of a walk through a garden on a spring day, the total absorption of a deep meditative state—these things and others like them constitute the reality of our experience of consciousness. Regardless of the content of any one of these experiences, no one in his or her right mind would doubt their reality. Any experience of consciousness—from the most mundane to the most elevated—has a certain coherence and, at the same time, a high degree of privacy, which means that it always exists from a particular point of view. The experience of consciousness is entirely subjective. The paradox, however, is that despite the indubitable reality of our subjectivity and thousands of years of philosophical examination, there is little consensus on what consciousness is. Science, with its characteristic third-person

6.

The Question of Consciousness

human existence or even, for that matter, to answer the question of the origination of life. This is not to deny that science does, and will continue to, have a lot to say about the origination of the tremendous diversity of life-forms. However, I do believe that, as a society, we must accept a degree of humility toward the limits of our scientific knowledge of ourselves and the world we live in.

If twentieth-century history—with its widespread belief in social Darwinism and the many terrible effects of trying to apply eugenics that resulted from it—has anything to teach us, it is that we humans have a dangerous tendency to turn the visions we construct of ourselves into self-fulfilling prophecies. The idea of the "survival of the fittest" has been misused to condone, and in some cases to justify, excesses of human greed and individualism and to ignore ethical models for relating to our fellow human beings in a more compassionate spirit. Thus, irrespective of our conceptions of science, given that science today occupies such an important seat of authority in human society, it is extremely important for those in the profession to be aware of their power and to appreciate their responsibility. Science must act as its own corrective to popular misconceptions and misappropriations of ideas that could have disastrous implications for the world and humanity at large.

Regardless of how persuasive the Darwinian account of the origins of life may be, as a Buddhist, I find it leaves one crucial area unexamined. This is the origin of sentience—the evolution of conscious beings who have the capacity to experience pain and pleasure. After all, from the Buddhist perspective, the human quest for knowledge and understanding of one's existence stems from a profound aspiration to seek happiness and overcome suffering. Until there is a credible understanding of the nature and origin of consciousness, the scientific story of the origins of life and the cosmos will not be complete.

There are explicitly cognitive states, like belief, memory, recognition, and attention on the one hand, and explicitly affective states, like the emotions on the other. In addition, there seems to be a category of mental states that function primarily as causal factors in that they motivate us into action. These include volition, will, desire, fear, and anger. Even within the cognitive states, we can draw distinctions between sensory perceptions, such as visual perception, which has a certain immediacy in relation to the objects being perceived, and conceptual thought processes, such as imagination or the subsequent recollection of a chosen object. These latter processes do not require the immediate presence of the perceived object, nor do they depend upon the active role of the senses.

In Buddhist philosophy of mind, we find discussion of various typologies of mental phenomena together with their distinct characteristics. First, there is the following sixfold typology: experiences of sight, hearing, smell, taste, touch, and the mental states. The first five are sensory experiences, while the last covers a wide range of mental states, from memory, will, and volition to imagination. The mental states dependent upon the five senses are thoroughly contingent upon sensory faculties that are understood to be material, while mental experiences enjoy a greater independence from physical bases.

One division of the Yogacara school posits two additions to this typology, making it eightfold. The exponents of this view argue that even mental perception is too transient and contingent to account for the profound unity we observe both in our subjective experience and in our sense of selfhood. They posit that, underlying all these fluctuating, contingent mental states, there must exist a basic mind that retains its integrity and continuum throughout the lifetime of an individual. This, they argue, is best understood as "foun-

dational consciousness," the basis of all mental phenomena. Inseparable from this foundational consciousness is the instinctual
thought "I am," a thought which Yogacara sees as a distinct stream
of consciousness.

The Middle Way school, which is the school whose standpoint
is generally recognized by Tibetan thinkers, including myself, to
represent the apex of Buddhist philosophical thinking, rejects this
typology and contends that the entire spectrum of consciousness is
adequately encompassed within the sixfold typology. In particular,
the Middle Way school is uncomfortable with the potentially essen-
tialist implications of "foundational consciousness" as postulated in
the eightfold system.

The question is, What defines this diversity of phenomena as
belonging to one family of experience, which we call "mental"? I remember most vividly my first lesson on epistemology as a child,
when I had to memorize the dictum "The definition of the mental
is that which is luminous and knowing." Drawing on earlier Indian
sources, Tibetan thinkers defined consciousness. It was years later
that I realized just how complicated is the philosophical problem
hidden behind this simple formulation. Today when I see nine-
year-old monks confidently citing this definition of consciousness
on the debating floor, which is such a central part of Tibetan
monastic education, I smile.

These two features—luminosity, or clarity, and knowing, or cog-
nizance—have come to characterize "the mental" in Indo-Tibetan
Buddhist thought. *Clarity* here refers to the ability of mental states
to reveal or reflect. *Knowing,* by contrast, refers to mental states' faculty to perceive or apprehend what appears. All phenomena possessed of these qualities count as mental. These features are
difficult to conceptualize, but then we are dealing with phenomena
that are subjective and internal rather than material objects that

may be measured in spatiotemporal terms. Perhaps it is because of these difficulties—the limits of language in dealing with the subjective—that many of the early Buddhist texts explain the nature of consciousness in terms of metaphors such as light or a flowing river. As the primary feature of light is to illuminate, so consciousness is said to illuminate its objects. Just as in light there is no categorical distinction between the illumination and that which illuminates, so in consciousness there is no real difference between the process of knowing or cognition and that which knows or cognizes. In consciousness, as in light, there is a quality of illumination.

In talking about mental phenomena which, according to Buddhist understanding, have the two defining characteristics of luminosity and knowing, there is a danger that one might assume Buddhism is proposing a version of Cartesian dualism—namely, that there are two independent substances, one called "matter" and the other called "mind." To allay any possible confusion, I feel a little digression on the basic classification of reality proposed in Buddhist philosophy is necessary. Buddhism suggests that there are three fundamentally distinct aspects or features of the world of conditioned things, the world in which we live:

1. Matter—physical objects
2. Mind—subjective experiences
3. Abstract composites—mental formations

As to what constitutes the world of matter, there is not much difference between Buddhist thought and modern science. Again, in defining the key characteristics of material phenomena, there would seem to be broad consensus between the two investigative traditions. We see properties—such as extension, spatiotemporal locality, and so on—as defining features of the material world. In addition to these manifestly material objects, from the Buddhist point of view, phenomena like subtle particles, the various fields (electromagnetic), and the forces of nature (gravity) belong to this first realm of reality. However, for Buddhist philosophers, reality is not exhausted by the contents of this realm.

There is also the realm of subjective experiences, such as our thought processes, sensory perceptions, sensations, and the rich tapestry of emotions. From the Buddhist perspective, much of this world can be found also in other sentient beings. Though heavily contingent upon a physical base—including neural networks, brain cells, and sensory faculties—the mental realm enjoys a status separate from the material world. From the Buddhist perspective, the mental realm cannot be reduced to the world of matter, though it may depend upon that world to function. With the exception of one materialist school in India, most ancient Indian and Tibetan philosophical schools agree on the impossibility of reducing the mental to a subset of the physical.

There is, moreover, a third realm of reality, the abstract composites, which can be characterized neither as physical in the sense of being composed of material constituents nor as mental in the sense of inner subjective experiences. By this I am referring to many features of reality that are integral to our understanding of the world. Phenomena such as time, concepts, and logical principles, which are essentially constructs of our mind, are distinct from the first two realms. Admittedly, all the phenomena that belong to this third world are contingent upon either the first or the second—that is physical or mental—domain of phenomena, yet they have characteristics distinct from the other two.

I gather that this taxonomy of reality, which goes back to the earliest phases of Buddhism's philosophical tradition, is almost identical to that proposed by Karl Popper. Popper called them "the first world," "the second world," and "the third world." By these he meant (1) the world of things or physical objects; (2) the world of subjective experiences, including thought processes; and (3) the world of statements in themselves—the content of thoughts as opposed to the mental process. It is striking that Popper, whom I know had no background in Buddhist thought, arrived at an almost identical classification of the categories of reality. Had I known this curious convergence between his thought and Buddhism in the times I met with Popper, I would certainly have pursued it with him.

Western philosophy and science have, on the whole, attempted to understand consciousness solely in terms of the functions of the brain. This approach effectively grounds the nature and existence of the mind in matter, in an ontologically reductionist manner. Some view the brain in terms of a computational model, comparing it to artificial intelligence; others attempt an evolutionary model for the emergence of the various aspects of consciousness. In modern neuroscience, there is a deep question about whether the mind and consciousness are any more than simply operations of the brain, whether sensations and emotions, are more than chemical reactions. To what extent does the world of subjective experience depend on the hardware and working order of the brain? It must to some significant extent, but does it do so entirely? What are the necessary and sufficient causes for the emergence of subjective mental experiences?

Many scientists, especially those in the discipline of neurobiology, assume that consciousness is a special kind of physical process that arises through the structure and dynamics of the brain. I vividly remember a discussion I had with some eminent neuroscientists at an American medical school. After they kindly showed me the latest scientific instruments to probe ever deeper into the human brain, such as MRI (magnetic resonance imaging) and ECG (electrocardiograph), and let me view a brain operation in progress (with the family's permission), we sat down to have a conversation on the current scientific understanding of consciousness. I said to one of the scientists: "It seems very evident that due to changes in the chemical processes of the brain, many of our subjective experiences like perception and sensation occur. Can one envision the reversal of this causal process? Can one postulate that pure thought itself could effect a change in the chemical processes of the brain?" I was asking whether, conceptually at least, we could allow the possibility of both upward and downward causation.

The scientist's response was quite surprising. He said that since all mental states arise from physical states, it is not possible for downward causation to occur. Although, out of politeness, I did not respond at the time, I thought then and still think that there is as yet no scientific basis for such a categorical claim. The view that all mental processes are necessarily physical processes is a metaphysical assumption, not a scientific fact. I feel that, in the spirit of scientific inquiry, it is critical that we allow the question to remain open, and not conflate our assumptions with empirical fact.

I notice that there is a group of scientists and philosophers who appear to believe that scientific thinking derived from quantum physics could provide an explanation of consciousness. I remember having some conversations with David Bohm about his idea of an "implicate order," in which both matter and consciousness manifest according to the same principles. Because of this shared nature, he contends, it is not surprising that we find a great similarity of order between thought and matter. Although I never fully understood Bohm's theory of consciousness, his emphasis on a holistic understanding of reality—including both mind and matter—suggests an avenue by which to look for a comprehensive understanding of the world.

In 2002 I met with a group of scientists at the University of Canberra in Australia on the subject of the unconscious mind. The astrophysicist Paul Davies argued that he could envision how one might formulate a quantum theory of consciousness. I must admit that every time a quantum explanation of consciousness is given, I become utterly lost. It is conceivable that quantum physics, with its logic-defying notions of nonlocality, the superposition of wave and particle properties, and Heisenberg's uncertainty principle, could offer deeper insights in specific areas of cognitive activity. Still, I can't quite see how a quantum theory of consciousness will fare better than a cognitive or a neurobiological explanation based on the classical understanding of physical processes in the brain. The only difference between the two explanations is the subtlety of the physical bases being correlated to the experience of consciousness. At least in my view, so long as the subjective experience of consciousness cannot be fully accounted for, the explanatory gap between the physical processes that occur in the brain and the processes of consciousness will remain as wide as ever.

Neurobiology has had tremendous success in mapping the brain and understanding its component parts. The process is fascinating and its results most interesting. Yet even here the part of the brain in which consciousness resides (if there is such a place) remains controversial. Some have suggested the cerebellum, some the reticular formation, some the hippocampus. Despite this lack of agreement, there seems to be a broadly shared assumption among neuroscientists that consciousness can be explained in terms of neurobiological processes.

Underlying this assumption is the confidence that all mental states, both cognitions and sensations, can be correlated to processes in the brain. With the invention of new powerful instruments, neuroscientists' knowledge of the correlation between various cognitive activities and brain processes has reached truly astonishing levels. For example, at one of the Mind and Life conferences, the psychologist Richard Davidson presented a detailed description of how many of the "negative" emotions, such as fear and hate, appear to be intimately associated with a part of the brain called the amygdala. The association between these emotional states and that part of the brain is so strong that in patients whose brains have been damaged in that region, emotions like fear and apprehension are said to be absent.

I remember making the observation that if experiments conclusively demonstrated that neutralizing that part of the brain would have no damaging consequences to the individual, then excising the amygdala could prove a most effective spiritual practice! Of course, the situation is not that simple. It turns out that, in addition to providing a neural basis for our negative emotions, the amygdala has other important roles to play, such as detection of danger, without which we would become incapacitated in many ways.

Despite the tremendous success in observing close correlations between parts of the brain and mental states, I do not think current neuroscience has any real explanation of consciousness itself. Neuroscience can probably tell us that when activity can be observed in this or that part of the brain the subject must be experiencing such and such a cognitive state. But it leaves open the question of why this is so. Furthermore, it does not and probably could not explain why, when such and such a brain activity occurs, the subject undergoes such and such an experience. For example, when a subject perceives the color blue, no amount of neurobiological explanation will get to the bottom of the experience. It will always leave out what it feels like to see blue. Similarly, a neuroscientist may be able to tell us whether a subject is dreaming, but can a neurobiological account explain the content of a dream?

A distinction can be made, however, between this as a methodological suggestion and the metaphysical assumption that mind is no more than a function or emergent property of matter. But assuming mind is reducible to matter leaves a huge explanatory gap. How do we explain the emergence of consciousness? What marks the transition from non-sentient to sentient beings? A model of increasing complexity based on evolution through natural selection is simply a descriptive hypothesis, a kind of euphemism for "mystery," and not a satisfactory explanation.

Crucial to understanding the Buddhist concept of consciousness—and its rejection of the reducibility of mind to matter—is its theory of causation. The issue of causality has long been a major focus of philosophical and contemplative analysis in Buddhism. Buddhism proposes two principal categories of cause. These are the "substantial cause" and the "contributory or complementary cause." Take the example of a clay pot. The substantial cause refers to the "stuff" that turns into a particular effect, namely, the clay that becomes the pot. By contrast, all the other factors that contribute toward bringing about the pot—such as the skill of the potter, the potter himself, and the kiln that fired the clay—remain complementary in that they make it possible for the clay to turn into the pot. This distinction between the substantial and the contributory cause of a given event or object is of the utmost importance for understanding the Buddhist theory of consciousness. According to Buddhism, though consciousness and matter can and do contribute toward the origination of each other, one can never become a substantial cause of the other.

In fact, it is on this premise that Buddhist thinkers like Dharmakirti have rationally argued for the tenability of the theory of rebirth. Dharmakirti's argument can be formulated as follows: The consciousness of the newborn infant comes about from a preceding instance of cognition, which is an instance of consciousness just like the present moment of consciousness.

The issue revolves around the argument that the various instances of consciousness we experience come into being because of the presence of preceding instances of consciousness; and since matter and consciousness have totally different natures, the first moment of consciousness of the new being must be preceded by its substantial cause, which must be a moment of consciousness. In this way, the existence of a previous life is affirmed.

Some other Buddhist thinkers, such as Bhavaviveka in the sixth century, have tried to argue for preexistence on the basis of habitual instincts, such as the newborn calf's instinctive knowledge of where to find its mother's teats and how to suck milk. These thinkers make the case that without assuming some form of preexistence, the phenomenon of "innate knowledge" cannot be coherently explained.

Regardless of how persuasive these arguments are, there are many examples of very young children with apparent memories of "previous lives," not to mention the numerous recollections of the Buddha's own past lives found in the scriptures. I know of a remarkable case of a young girl from Kanpur in the Indian state of Uttar Pradesh in the early 1970s. Although initially her parents dismissed the girl's descriptions of a second set of parents in a place she described specifically, the girl's accounts were so concrete that they began to take her seriously. When the two whom she claimed to be her parents during her previous life came to see her, she told them very specific details of their deceased child's life, which only a close member of the family could have known. As a result, when

I met her, the other two parents had also fully embraced her as a member of their family. This is only anecdotal evidence, but such phenomena cannot be easily dismissed.

Reams have been written on the analysis of this form of Buddhist reasoning, the technical aspects of which lie outside the scope of the present discussion. The point I wish to make is that Dharmakirti clearly did not think that the theory of rebirth was purely a matter of faith. He felt that it falls within the purview of what he characterized as "slightly hidden" phenomena, which can be verified by means of inference.

A crucial point about the study of consciousness as opposed to the study of the physical world relates to the personal perspective of accounts like this. In examining the physical world, leaving aside the problematic issue of quantum mechanics, we are dealing with phenomena that lend themselves well to the dominant scientific method of the objective, third-person method of inquiry. On the whole, we have a sense that a scientific explanation of the physical world does not exclude the key elements of the field being described. In the realm of subjective experiences, however, the story is completely different. When we listen to a purely third-person, "objective" account of mental states, whether it is a cognitive psychological theory, a neurobiological account, or an evolutionary theory, we feel that a crucial dimension of the subject has been left out. I am referring to the phenomenological aspect of mental phenomena, namely the subjective experience of the individual.

Even from this brief discussion, it is, I think, clear that the third-person method—which has served science so well in so many areas—is inadequate to the explanation of consciousness. What is required, if science is successfully to probe the nature of consciousness, is nothing short of a paradigm shift. That is, the third-person perspective, which can measure phenomena from the point of view of an independent observer, must be integrated with a first-person perspective, which will allow the incorporation of subjectivity and the qualities that characterize the experience of consciousness. I am suggesting the need for the method of our investigation to be appropriate to the object of inquiry. Given that one of the primary characteristics of consciousness is its subjective and experiential nature, any systematic study of it must adopt a method that will give access to the dimensions of subjectivity and experience.

A comprehensive scientific study of consciousness must therefore embrace both third-person and first-person methods: it cannot ignore the phenomenological reality of subjective experience but must observe all the rules of scientific rigor. So the critical question is this: Can we envision a scientific methodology for the study of consciousness whereby a robust first-person method, which does full justice to the phenomenology of experience, can be combined with the objectivist perspective of the study of the brain?

Here I feel a close collaboration between modern science and the contemplative traditions, such as Buddhism, could prove beneficial. Buddhism has a long history of investigation into the nature of mind and its various aspects—this is effectively what Buddhist meditation and its critical analysis constitute. Unlike that of modern science, Buddhism's approach has been primarily from first-person experience. The contemplative method, as developed by Buddhism, is an empirical use of introspection, sustained by rigorous training in technique and robust testing of the reliability of experience. All meditatively valid subjective experiences must be verifiable both through repetition by the same practitioner and through other individuals being able to attain the same state by the same practice. If they are thus verified, such states may be taken to be universal, at any rate for human beings.

The Buddhist understanding of mind is primarily derived from empirical observations grounded in the phenomenology of experience, which includes the contemplative techniques of meditation. Working models of the mind and its various aspects and functions are generated on this basis; they are then subjected to sustained critical and philosophical analysis and empirical testing through both meditation and mindful observation. If we want to observe how our perceptions work, we may train our mind in attention and learn to observe the rising and falling of perceptual processes on a moment-by-moment basis. This is an empirical process which results in firsthand knowledge of a certain aspect of how the mind works. We may use that knowledge to reduce the effects of emotions such as anger or resentment (indeed, meditation practitioners in search of overcoming mental affliction would wish to do this), but my point here is that this process offers a first-person empirical method with relation to the mind.

I am aware that there is a deep suspicion of first-person methods in modern science. I have been told that, given the problem inherent in developing objective criteria to adjudicate between competing first-person claims of different individuals, introspection as a method for the study of the mind in psychology has been abandoned in the West. Given the dominance of third-person scientific method as a paradigm for acquiring knowledge, this disquiet is entirely understandable.

I agree with the Harvard psychologist Stephen Kosslyn, who has conducted pioneering research into the role of introspection in imagination; he argued at a recent Mind and Life conference, "Investigating the Mind," that it is critical to recognize the natural boundaries of introspection. He argued that, no matter how highly trained a person may be, we have no evidence that his or her introspection can reveal the intricacies of the neural networks and the

biochemical composition of the human brain, or the physical correlates of specific mental activities—tasks that can be most accurately performed by empirical observation through application of
powerful instruments. However, a disciplined use of introspection
would be most suited to probe the psychological and phenomenological aspects of our cognitive and emotional states.

What occurs during meditative contemplation in a tradition
such as Buddhism and what occurs during introspection in the ordinary sense are two quite different things. In the context of Buddhism, introspection is employed with careful attention to the
dangers of extreme subjectivism—such as fantasies and delu-
sions—and with the cultivation of a disciplined state of mind. Refinement of attention, in terms of stability and vividness, is a
crucial preparation for the utilization of rigorous introspection,
much as a telescope is crucial for the detailed examination of celestial phenomena. Just as in science, there is a series of protocols
and procedures which contemplative introspection must employ.
Upon entering a laboratory, someone untrained in science would
not know what to look at, would have no capacity to recognize
when something is found; in the same way, an untrained mind will
have no ability to apply the introspective focus on a chosen object
and will fail to recognize when processes of the mind show themselves. Just like a trained scientist, a disciplined mind will have the
knowledge of what to look for and the ability to recognize when
discoveries are made.

It may well be that the question of whether consciousness can
ultimately be reduced to physical processes, or whether our subjective experiences are non-material features of the world, will remain a matter of philosophical choice. The key issue here is to
bracket out the metaphysical questions about mind and matter,
and to explore together how to understand scientifically the vari-

ous modalities of the mind. I believe that it is possible for Buddhism and modern science to engage in collaborative research in the understanding of consciousness while leaving aside the philosophical question of whether consciousness is ultimately physical. By bringing together these two modes of inquiry, both disciplines may be enriched. Such collaborative study will contribute not only to greater human understanding of consciousness but also to a better understanding of the dynamics of the human mind and its relation to suffering. This is a precious gateway into the alleviation of suffering, which I believe to be our principal task on this earth.

7.

Toward a Science
of Consciousness

In order for the study of consciousness to be complete, we need a methodology that would account not only for what is occurring at the neurological and biochemical levels but also for the subjective experience of consciousness itself. Even when combined, neuroscience and behavioral psychology do not shed enough light on the subjective experience, as both approaches still place primary importance on the objective, third-person perspective. Contemplative traditions on the whole have historically emphasized subjective, first-person investigation of the nature and functions of consciousness, by training the mind to focus in a disciplined way on its own internal states.

In this kind of analysis the observer, the object, and the means of investigation are all aspects of the same thing, namely the mind of the individual experimenter. In Buddhism, this mental training is called *bhavana,* which is usually translated as "meditation" in

English. The original Sanskrit term *bhavana* carries connotations of cultivation, in the sense of cultivating a habit, while the Tibetan term, *gom,* literally means "to familiarize." So the idea is a disciplined mental practice of cultivating familiarity with a given object, whether an external object or an internal experience.

People often understand *meditation* to refer simply to an emptying of the mind, or a relaxation practice, but that is not what I mean here. The practice of *gom* does not imply any mysterious or mystical state or ecstasy open only to a few gifted individuals. Nor does it entail non-thinking or the absence of mental activity. The term *gom* refers both to a means, or a process, and to a state that may arise as a result of the process. I am concerned here primarily with *gom* as a means, which implies a rigorous, focused, and disciplined use of introspection and mindfulness to probe deeply into the nature of a chosen object. From the scientific point of view, this process can be compared with rigorous empirical observation.

The difference between science as it stands now and the Buddhist investigative tradition lies in the dominance of the third-person, objective method in science and the refinement and utilization of first-person, introspective methods in Buddhist contemplation. In my view, the combination of the first-person method with the third-person method offers the promise of a real advance in the scientific study of consciousness. A great deal can be accomplished by the third-person method. As brain-imaging technologies become ever more effective, it is possible to observe closely the physical correlates of our rich world of subjective experience — such as neural connections, biochemical changes, the locations in the brain associated with specific mental activities, and the temporal processes (often at the minute level of milliseconds) by which the brain responds to external stimuli. I had the pleasure of seeing this firsthand when I visited Richard Davidson's laboratory at the University of Wisconsin in Madison during the spring of 2001.

This is a brand-new lab with state-of-the-art brain-imaging technology and instruments. Davidson has a young and exciting group of colleagues there, and one of his projects—of special interest to me—is an ongoing series of experiments on meditators. He showed me around and demonstrated the different machines. There was an EEG (electroencephalograph), primarily used to detect electrical activity in the brain. This is made like a cap to fit on the head with many sensors attached to it, and apparently the one in Davidson's lab, with 256 sensors, is among the most sophisticated in the world. In addition, there was an MRI (magnetic resonance imaging) machine, which is so sensitive that the subject has to be very still while inside for an accurate reading to take place. The strength of the EEG, I am told, is its speed (astoundingly, it can detect changes in the brain within a thousandth of a second), while the power of the MRI lies in its ability to pinpoint locations of brain activity to within a millimeter.

The day before my visit they had used these machines in a detailed experiment involving an experienced meditator whom I have known for a long time performing a variety of meditation practices. Davidson showed me a computer screen with many scanned images of the subject's brain, with different colors indicating differing types of activity.

On the following day, we held a formal meeting where Davidson presented the preliminary results of his studies. The psychologist Paul Ekman joined the discussions and gave a preliminary report on his ongoing work on a large number of groups of subjects, including meditators. Scientific experimentation on meditators has quite a long history now, going back to the experiments conducted by Herbert Benson of the Harvard Medical School in the 1980s. Benson monitored the physiological changes in body heat and oxygen consumption of meditators performing *tummo* practice, which involves among other things the generation of heat at a specific point in the body. Like Benson, Richard Davidson's group has undertaken experiments on hermits in the Himalayas, including in the mountains around Dharamsala. Because conducting experiments in the mountains necessitates the use of mobile equipment, this work is bound to be limited, at least until mobile technology catches up.

Scientific experimentation on human subjects raises numerous ethical issues, a problem which the scientific community takes most seriously. For the hermits who have chosen a life of solitude in the mountains, there is the added complication that such experimentation constitutes a profound intrusion into their lives and spiritual practice. It is not surprising that initially many were reluctant. Apart from anything else, most simply couldn't see the point, other than satisfying the curiosity of some odd men carrying machines. However, I felt very strongly (and still do feel) that the application of science to understanding the consciousness of meditators is most important, and I made a great effort to persuade the hermits to allow the experiments to take place. I argued that they should undergo the experiments out of altruism; if the good effects of quieting the mind and cultivating wholesome mental states can be demonstrated scientifically, this may have beneficial results for others. I only hope I was not too heavy-handed. A number of hermits accepted, persuaded, I hope, by my argument rather than simply submitting to the authority of the Dalai Lama's office.

All this work can illuminate one side of the picture of consciousness. But unlike the study of a three-dimensional material object in space, the study of consciousness, including the entire range of its phenomena and everything that falls under the rubric of subjective experience, has two components. One is what happens to the brain and to the behavior of the individual (what brain science and behavioral psychology are equipped to explore), but the other is the phenomenological experience of the cognitive, emotional, and psychological states themselves. It is for this latter element that the application of a first-person method is essential. To put it another way, although the experience of happiness may coincide with certain chemical reactions in the brain, such as an increase in serotonin, no amount of biochemical and neurobiological description of this brain change can explain what happiness is.

While the Buddhist contemplative tradition has not had access to scientific means of gaining insight into the brain processes, it has an acute understanding of the mind's capacity for transformation and adaptation. Until recently, I gather, scientists believed that after adolescence the hardware of the human brain becomes relatively unchangeable. But new discoveries in neurobiology have uncovered a remarkable potential for changeability in the human brain even in adults as old as I am. At the Mind and Life conference in Dharamsala in 2004, I learned of the growing subdiscipline of neuroscience dealing with this question, called "brain plasticity." This phenomenon suggests to me that traits that were assumed to be fixed—such as personality, disposition, even moods—are not permanent, and that mental exercises or changes in the environment can affect these traits. Already experiments have shown that experienced meditators have more activity in the left frontal lobe, the part of the brain associated with positive emotions, such as happiness, joy, and contentment. These findings imply that happiness is something we can cultivate deliberately through mental training that affects the brain.

The seventh-century philosopher-monk Dharmakirti presents

a sophisticated argument in support of the proposition that, through disciplined meditative training, substantive changes can be effected in human consciousness, including the emotions. A key premise underlying his argument is the universal law of cause and effect, which suggests that the conditions affecting the cause have an inevitable impact on the result. This principle is very ancient in Buddhism—the Buddha himself argued that if one wishes to avoid certain types of results, one needs to change the conditions that give rise to them. So, if one changes the conditions of one's state of mind (which normally give rise to particular habitual patterns of mental activity), one can change the traits of one's consciousness and the resulting attitudes and emotions.

The second key premise is the universal law of impermanence, which was part of many of the Buddha's earliest teachings. This law states that all conditioned things and events are in constant flux. Nothing—not even in the material world, which we tend to perceive as enduring—remains static or permanent. So this law suggests that anything produced by causes is susceptible to change—and if one creates the right conditions, one can consciously direct such change to a transformation of the state of one's mind.

Like other Buddhist thinkers before him, Dharmakirti invokes what could be called a "psychological law" in that he sees various psychological states, including the emotions, as a field of forces in which opposing families of mental states interact in a constant dynamic. Within the domain of the emotions, there might be a family consisting of hate, anger, hostility, and so forth, while in opposition is a family of positive emotions, like love, compassion, and empathy. Dharmakirti argues that if one side of any such polarity is stronger, the other is weaker in any given individual at any given time. So if one works to increase, reinforce, and strengthen the positive groups, one will correspondingly weaken the negative ones, thus effectively bringing about transformation in one's thoughts and emotions.

Dharmakirti illustrates the complexity of this process by a series of vivid analogies from everyday experience. The opposing forces might be seen as heat and cold, which can never coexist without one undermining the other but at the same time neither can eliminate the other instantaneously—the process is gradual. Probably Dharmakirti had in mind the effect of lighting a fire to warm a cold room or rainfall in the monsoon cooling the tropics, where he lived. By contrast, Dharmakirti talks of the light of a lamp as immediately dispelling darkness.

This law whereby two opposing states cannot coexist without one undermining the other is the key premise in the Buddhist argument for the transformability of consciousness—it means that the cultivation of loving-kindness can over a period diminish the force of hate in the mind. Further, Dharmakirti argues that the removal of a basic condition will remove its effects. So that, by eliminating the cold, for example, one effectively removes all its attendant results, such as goose pimples, shivering, and chattering teeth.

Dharmakirti goes even further and suggests that, unlike physical abilities, the qualities of the mind have the potential for limitless development. Contrasting mental training with the physical training of athletes, especially long jumpers, he argues that in athletic prowess, although there might be a wide range of levels to which individual athletes can aspire, there is a fundamental limit imposed by the nature and constitution of the human body, no matter how much training one may undergo or how outstanding a natural athlete one may be. Even the illegal use of drugs in modern athletics, which may extend the body's limits marginally, cannot in fact push the human body beyond the fundamental limitations of its own nature. By contrast, Dharmakirti argues that the natural constraints on consciousness are far fewer and are removable, so that in principle it is possible for a mental quality like compassion to be developed to a limitless degree. In fact, for Dharmakirti, the greatness of the Buddha as a spiritual teacher lies not so much in his mastery of various fields of knowledge as in his having attained the perfection of boundless compassion for all beings.

Even before Dharmakirti, there was a widespread understanding within Indian Buddhism of the mind's capacity for transformation from a negative state to a state of tranquil and wholesome purity. A Mahayana work of the fourth century, *The Sublime Continuum,* which is attributed to Maitreya, and a shorter work attributed to Nagarjuna entitled *Praise to the Ultimate Expanse,* argue that the essential nature of the mind is pure and that its defilements are removable through meditative purification. These treatises themselves draw on the notion of Buddha nature, the natural potential for perfection that lies in all sentient beings (including animals). *The Sublime Continuum* and Nagarjuna's *Praise* offer two principal theses for the basic transformability of mind toward a positive end. The first is the conviction that all negative traits of the mind may be purified by applying the appropriate antidotes. This means that the pollutants of the mind are not seen as essential or intrinsic to it and that the mind's essential nature is pure. From the scientific point of view, these are metaphysical assumptions. The second is that the capacity for positive transformation lies naturally within the constitution of the mind itself—which follows the first thesis.

The texts on Buddha nature employ metaphors to illustrate the theme of the innate purity of mind's essential nature. Nagarjuna's

Praise to the Ultimate Expanse opens with a series of vivid images which contrast the mind's essential purity with its pollutants and afflications. Nagarjuna likens this natural purity to the butter lying unextracted in unchurned milk, to an oil lamp concealed inside a vase, to a pristine deposit of lapis lazuli buried in a rock, and to a seed covered by its husk. When the milk is churned, the butter is revealed; holes may be made in the vase so that the lamp's illumination is released; when the gem is dug out, the brilliance of lapis lazuli shines forth; when the husk is removed, the seed can germinate. In the same manner, when our afflictions are cleansed through the sustained cultivation of insight penetrating the ultimate nature of reality, the innate purity of the mind—which Nagarjuna calls the "ultimate expanse"—becomes manifest.

The *Praise to the Ultimate Expanse* goes a step further and asserts that just as subterranean water retains its purity as water, so even within the afflictions the perfected wisdom of an enlightened mind can be found. *The Sublime Continuum* describes the obscuration of our mind's natural purity with analogies to a Buddha sitting on a soiled lotus, honey concealed within a beehive, a piece of gold dropped in filth, precious treasure buried beneath a beggar's home, the potential matured plant in a young shoot, and an image of the Buddha hidden inside a rag.

For me, these two Indian Buddhist classics and the various works that belong to the same genre, which are written in highly evocative and poetic language, were a refreshing contrast to the rigorously logical and systematic writings that are part of Buddhist philosophical tradition. For Buddhists, the theory of Buddha nature—the notion that the natural capacity for perfectibility lies within each of us—is a deeply and continually inspiring concept.

My point here is not to suggest that we could use the scientific method to prove the validity of the theory of Buddha nature but simply to show some of the ways in which the Buddhist tradition has attempted to conceptualize the transformation of consciousness. Buddhism has long had a theory of what in neuroscience is called the "plasticity of the brain." The Buddhist terms in which this concept is couched are radically different from those used by cognitive science, but what is significant is that both perceive consciousness as highly amenable to change. The concept of neuroplasticity suggests that the brain is highly malleable and is subject to continual change as a result of experience, so that new connections between neurons may be formed or even brand-new neurons generated. Research in this area specifically includes work on virtuosos—athletes, chess players, and musicians—whose intense training has been shown to result in observable changes in the brain. These kinds of subjects are interestingly parallel to skilled meditators, who are also virtuosos, and whose dedication to their practice involves a similar commitment of time and effort.

Whether we talk of the transformation of consciousness or of the introspective empirical analysis of what occurs in the mind, the observer needs a range of skills, carefully honed through repetition and training, and applied in a rigorous and disciplined manner. All these practices assume a certain ability to direct one's mind to a chosen object and to hold the attention there for a period, however brief. An assumption is also made that, through constant habituation, the mind learns to improve the quality of whatever faculty is being primarily applied, whether it is attention, reasoning, or imagination. The understanding is that through such prolonged and regular practice, the ability to perform the exercise will become almost second nature. Here the parallel with athletes or musicians is very clear, but one might equally think of learning how to swim or how to ride a bicycle. Initially, these are very difficult, seemingly unnatural activities, but once you master the skill, they come quite easily.

One of the most basic mental trainings is the cultivation of mindfulness, especially performed on the basis of observing one's breath. Mindfulness is essential if one is to become consciously aware in a disciplined manner of whatever phenomena may occur within the mind or one's immediate environment. In our normal state, our mind remains unfocused for most of the time and our thoughts move from one object to another in a random and dissipated manner. By cultivating mindfulness, we learn first to become aware of this process of dissipation, so that we can gently fine-tune the mind to follow a more directed path toward the objects on which we wish to focus. Traditionally, the breath is seen as an ideal instrument for the practice of mindfulness. The great advantage of choosing one's breath as the object of mindfulness training is that breathing is an instinctive and effortless activity, something which we do as long we are alive, so there is no need to strive hard to find the object of this practice. In its developed form, mindfulness also brings about a highly refined sensitivity to everything that happens, however minute, in one's immediate vicinity and in one's mind.

One of the most crucial elements of training in mindfulness is the development and application of attention. Given that a significant percentage of children suffer from attention deficit problems in today's world, especially in more materially affluent societies, I am told that substantial efforts are being made to understand the faculty of attention and its causal dynamics. Here Buddhism's long history of training attention could make a contribution. In Buddhist psychology, *attention* is defined as the faculty that helps direct the mind to a chosen object among the variety of sensory information we experience in any given moment. We shall not be concerned here with the complex theoretical issues surrounding what attention is exactly—whether it is a single mechanism or of several types, or whether it is the same as controlled application of thought. Rather, let us take attention as a deliberate intention that helps us select a specific aspect or a characteristic of an object. The continued, voluntary application of attention is what helps us maintain a sustained focus on the chosen object.

Training in attention is closely linked with learning how to control our mental processes. I am sure most young people today, even many who have been diagnosed as suffering from attention deficit disorders, can enjoy a gripping film without distraction. Their problem is the ability to direct their attention willfully when there is more than one thing happening. Another factor has to do with habit. The less our familiarity, the greater the need for effort and deliberate application both to direct our attention to and to keep it on a chosen object or task. However, through habituation acquired in training, we become less dependent upon such deliberate effort. We know from personal experience that, through training, even tasks that seem extremely hard initially can become almost automatic. Buddhist psychology understands that through sustained, disciplined practice one's application of attention, which initially involves a great deal of effort, subsequently gives way to a limited mastery in which some effort is still required, and finally the task becomes effortless and spontaneous.

Another practice for the development of attention is single-pointed concentration. Here the observer may choose any kind of object, external or internal, but something that he or she can easily conjure the image of. The training proceeds with the deliberate placement of one's attention on the chosen object and the attempt to hold that attention as long as possible. This practice involves primarily the use of two faculties, mindfulness (which keeps the mind tied to the object) and introspective vigilance, which discerns whether distraction occurs in the mind and whether the vividness of the mind's focus has become lax. At the heart of this practice lies the development of two qualities of the disciplined mind—the stability of prolonged attention and the clarity or vividness with which the mind can perceive the object. In addition, the practitioner needs to learn to maintain equanimity, so that he or she does not apply excessive introspection onto the object, which would distort the object or destabilize mental composure.

When the practitioner notices as a result of introspection that he has become distracted, he needs to bring the mind back gently to the object. Initially, the time lapse between one's mind being distracted and detecting this distraction may be relatively lengthy, but after regular training, it will become shorter and shorter. In its developed form, this practice allows an observer to rest for long periods on the chosen object, noticing any changes that may occur, whether in the object or in the mind. Furthermore, the practitioner is said to have achieved a quality of mental pliancy, in that the mind has become easily serviceable and can be directed freely to any object. This state is described as the attainment of tranquil abiding of the mind (*shamatha* in Sanskrit, *shi ne* in Tibetan).

There are claims in the Buddhist meditation texts that a skilled practitioner can master this technique to such a point that she can hold her attention unwaveringly for four hours at a time. I knew a Tibetan meditator who was reputed to have achieved this state. Unfortunately, he has passed away; otherwise it would have been most interesting to examine him while in this state with all the sophisticated machinery in Richard Davidson's lab. One fruitful area of study for the emergent field of attention studies in Western psychology would be testing cases like this against the current scientific understanding of human attention, which I believe sets the maximum span at no more than a few minutes.

These meditative practices provide a settled and disciplined state of mind, but if our aim is to delve deeper into the subject under investigation, it is not adequate simply to have a focused mind. We must acquire the skill of probing the nature and characteristics of the object of our observation with as much precision as possible. This second-level training is known in the Buddhist literature as insight (*vipashyana* in Sanskrit, *lhak thong* in Tibetan). In tranquil abiding the emphasis is on holding one's focus without distraction, and single-pointedness is the key quality being sought. In insight the emphasis is on discerning investigation and analysis while maintaining one-pointedness without distraction.

In his classic *Stages of Meditation*, the eighth-century Indian Buddhist master Kamalashila gives a detailed account of how both tranquil abiding and insight may be systematically cultivated. These are combined so that they can be applied to deepening one's understanding of specific features of reality, to the point that one's understanding affects thoughts, emotions, and behavior. He particularly emphasizes the need for maintaining a fine equilibrium between the single-pointed placement of the mind on the one hand and the application of a focused beam of analysis on the other. This is because they are different mental processes with the potential of undermining each other. Single-pointed absorption on a chosen object requires holding the mind on the object with little movement and a kind of fusion, while insight requires a certain directed activity in which the mind moves from one aspect of the object to another.

When we are cultivating insight, Kamalashila advises that we begin investigation with as much sharpness of inquiry as possible and that we then try to hold the mind single-pointedly on the resultant insight for as long as possible. When the practitioner begins to lose the force of the insight, he advises recommencing the analytic process. This alternation may then lead to a higher level of mental capability, at which both analysis and absorption become relatively effortless.

As in any other discipline, tools help the experimenter focus his or her exploration. Since subjective experience can easily become derailed by fantasy or delusion, meditative tools such as structured analysis have been developed to focus contemplative exploration. Often, topics for analysis are prescribed. A meditator may choose among many topics to focus on. One of them is the transient nature of our own existence. Impermanence is chosen as a worthy object of meditation in Buddhism because, although we may understand it intellectually, we mostly do not behave as though we have integrated this awareness. A combination of analysis and concentration on this topic brings the insight to life so that we appreciate the preciousness of every moment of our existence.

To begin we become mindful of the body and the breath in a state of calm, and we cultivate awareness of the very subtle changes that occur in the mind and in the body during a period of practice, even between the in-breath and the out-breath. In this way, an experiential awareness arises that nothing within one's existence stays static or unchanging. As one fine-tunes this practice, one's awareness of change becomes ever more minute and dynamic. For example, one approach is to contemplate the complex web of circumstances that keep us alive, which leads to a deeper appreciation of the fragility of our continued existence. Another approach is a more graphic examination of bodily processes and functioning, particularly aging and decay. If a meditator had a deep knowledge of biology, then it is conceivable that there would be a specially rich content to his or her experience of this practice.

These thought experiments have been performed repeatedly over many centuries, and the results have been confirmed by thousands of great meditators. Buddhist practices are tested for efficacy and confirmed by reliable minds before they become tools for meditation to use.

If our purpose is to incorporate first-person perspectives into the scientific method in order to develop a means for studying consciousness, fortunately we don't need to keep up this practice perfectly for four hours at a stretch. What is needed is some degree of combination of the two techniques—single-pointedness and investigation. Disciplined training is the key. A physicist needs to go through training which includes skills such as mathematics, the ability to use various instruments, the critical faculty to know whether an experiment is correctly designed and whether the results support the hypothesis, as well as the expertise to interpret the results of past experiments. These skills can be acquired and developed only over a long period. Someone who wishes to learn the skills of the first-person method needs to devote a comparable amount of time and effort. It is important to stress here that, like the training of a physicist, the acquisition of mental skills is a matter of volition and focused effort; it is not a special mystical gift given to the few.

There are many other forms of meditation in the Buddhist tradition, including a vast body of practices that involve the use and enhancement of visualization and imagination, and various techniques for manipulating the vital energies in the body to induce progressively deeper and subtler states of mind, which are characterized by increasing freedom from conceptual elaboration. These states and practices may be an interesting area of scientific research and experimentation, in that they may suggest unexpected capacities and potentials within the human mind.

One area of possible research on meditation could be what the Tibetan tradition describes as the experience of the clear light state. This is a state of consciousness understood to be extremely subtle that manifests briefly in all human beings at the moment of death. Very brief similitudes of this state may occur naturally at other times, such as during sneezing, fainting, deep sleep, and sexual climax. The principal characteristic of the state is a total spontaneity, the absence of self-consciousness or self-grasping. In an experienced practitioner, this state can be deliberately induced through meditative techniques, and when it naturally occurs at death such an individual can sustain the state while maintaining mindfulness for a long period.

My own teacher Ling Rinpoche remained in the clear light of death for thirteen days; although he was clinically dead and had stopped breathing, he stayed in the meditation posture and his body showed no sign of decomposition. Another realized meditator remained in this state for seventeen days in the tropical heat of high summer in eastern India. It would be most interesting to know what was happening at the physiological level during this period, and if there might be any detectable signs at the biochemical level. When Richard Davidson's team came to Dharamsala, they were keen to do some experiments on this phenomenon, but when they were there — I am not sure whether I should say fortunately or unfortunately — no meditators died.

However, from the perspective of contributing to the emergence of a scientific method grounded in a rigorous first-person approach, these types of practice are not strictly relevant. In training ourselves to take consciousness itself as the object of first-person investigation, we must first stabilize the mind. The experience of attending to the mere present is a very helpful practice. The focus of this practice is a sustained training to cultivate the ability to hold the mind undistractedly on the immediate, subjective experience of consciousness. This is done as follows.

Before one begins formal sitting meditation, one develops a deliberate intention not to allow the mind to be distracted either by recollections of past experience or by hopes, anticipations, and fears about future events. This is done by making a silent pledge that during this meditation session the mind will not be seduced by thoughts of the past or the future and that it will remain fully focused on awareness of the present. This is critical because in our everyday normal states we tend to be tied either to recollections and vestiges of the past or to hopes and fears about the future. We tend to live either in the past or in the future and very rarely fully in the present. When one is actually in the meditation session, it may be helpful to face a wall that has no contrasting colors or patterns that might distract the attention. A muted color like cream or beige can be suitable, for these help create a simple background. When one is actually in the session, it is critical not to apply any exertion. Rather, one must simply observe the mind resting naturally in its own state.

As one sits, one will begin to notice that all sorts of thoughts arise in the mind, like a bubbling spring of never-ending internal chatter or the bustle of endless traffic. One should allow whatever thoughts arise to do so freely, regardless of whether one perceives them as wholesome or unwholesome. Do not reinforce them or repress them or subject them to evaluative judgment. Any of these responses will create further proliferation of thought, for it will provide the fuel that keeps the chain reaction going. One must simply observe the thoughts. When one does this, just as bubbles arise and dissolve into water, the discursive thought processes simply arise and dissolve within the mind.

Gradually, in the midst of the internal chatter, one will begin to glimpse what feels like a mere absence, a state of mind with no specific, determinable content. At the beginning, such states may be only fleeting experiences. Nevertheless, as one becomes more experienced in this practice, one will be able to prolong the intervals in one's normal proliferation of thought. Once this happens, there is a real opportunity to understand experientially what is described in the Buddhist definition of consciousness as "luminous and knowing." In this way a meditator will gradually be able to "grasp" the basic experience of consciousness and take that as the object of meditative investigation.

Consciousness is a very elusive object, and in this sense it is quite unlike the focus on a material object, such as biochemical processes. Yet its elusiveness may be compared with that of some objects of physics and biology, like subatomic particles or genes. Now that the methods and protocols for their investigation are fully established, these things seem familiar and even relatively uncontentious. All these studies are observation-driven, in that—regardless of the philosophical views scientists may bring to any given experiment—in the final analysis, it is the empirical observation based on evidence and the discovery of phenomena that must determine what is true. Likewise, whatever our philosophical views about the nature of consciousness, whether it is ultimately material or not, through a rigorous first-person method we can learn to observe the phenomena, including their characteristics and causal dynamics.

On this basis, I envision the possibility of broadening the scope of the science of consciousness and enriching our collective understanding of the human mind in scientific terms. Francisco Varela once told me that a European philosopher, Edmund Husserl, already suggested a similar approach to the study of consciousness. He described the method of proceeding from one's felt experience, without bringing into one's examination the extra dimension of metaphysical assumptions, as the act of "bracketing out metaphysics" from a phenomenological inquiry. This does not mean that the individual does not subscribe to a philosophical position but rather that he or she deliberately suspends personal beliefs for purposes of the analysis. In effect, something similar to bracketing is already at play in modern science.

Biology, for instance, has made tremendous advances in giving us a scientific understanding of life and its various forms and constituents, despite the fact that the conceptual and philosophical question of what life is remains open. Likewise, the remarkable feats of physics (especially in quantum mechanics) have been achieved without a clear answer to the question What is reality? and while many conceptual issues pertaining to their interpretation remain unresolved.

To some degree I think that experience of, indeed training in, some of these techniques of mental discipline (or others like them) will have to become an integral part of the training of the cognitive scientist if science is serious about gaining access to the full range of methods necessary for a comprehensive study of consciousness. Indeed, I would agree with Varela that if the scientific study of consciousness is ever to grow to full maturity—given that subjectivity is a primary element of consciousness—it will have to incorporate a fully developed and rigorous methodology of first-person empiricism. It is in this area that I feel there is a tremendous potential for established contemplative traditions, such as Buddhism, to make a substantive contribution to the enrichment of science and its methods. Moreover, there may well be substantive resources in the West's own philosophical tradition to help modern science develop its methods toward an accommodation of first-person perspectives. In this way, we may be able to expand our horizons toward a greater understanding of one of the key qualities that characterize our human existence, namely consciousness.

8

The Spectrum of Consciousness

In the emerging science of consciousness and the investigation of the mind and its various modalities, Buddhism and cognitive science take different approaches. Cognitive science addresses this study primarily on the basis of neurobiological structures and the biochemical functions of the brain, while Buddhist investigation of consciousness operates primarily from what could be called a first-person perspective. A dialogue between the two could open up a new way of investigating consciousness. The core approach of Buddhist psychology involves a combination of meditative contemplation, which can be described as a phenomenological inquiry; empirical observation of motivation, as manifested through emotions, thought patterns and behavior, and critical philosophical analysis.

The principal aim of Buddhist psychology is not to catalog the mind's makeup or even to describe how the mind functions; rather its fundamental concern is to overcome suffering, especially psychological and emotional afflictions, and to clear those afflictions. In classical Buddhist sources, there are three separate disciplines for the study of consciousness. The Abhidharma focuses on examination of the causal processes of the hundreds of mental and emotional states, our subjective experience of these states, and their effects on our thoughts and behavior. It is related to what could be called psychology (including cognitive therapy) and phenomenology. Second, Buddhist epistemology analyzes the nature and characteristics of perception, knowledge, and the relationship between language and thought in order to develop a conceptual framework for understanding the various aspects of consciousness—thoughts, emotions, and so on. Finally, Vajrayana uses visualization, thoughts, emotions, and various physical techniques (such as yogic exercises) in an intense meditative effort to accentuate wholesome ways of being and to transmute the afflictions of the mind. It is concerned not with discovering an independent permanent entity called "the mind" but rather with understanding the nature of the ordinary mind and effecting its transformation into a non-afflicted, clearer state.

The Buddhist approach to the study of consciousness is based on the understanding of functions and the modalities of the mind and their causal dynamics—and this is precisely an area where Buddhist understanding can most readily intersect with a scientific approach because, like that of science, much of the Buddhist investigation of consciousness is empirically based.

I began to learn about the various aspects of the mind as part of my introduction to what is called *lo rig*, which literally means "awareness and intelligence." This topic is taught to trainee monks—normally at the age of nine or ten, after the novitiate ordination at the age of eight. First, my tutors—primarily Ling Rinpoche at this time—made me memorize a working definition of the nature of mental events and the main categories of cognitive and emotional states. Although I did not have a clear idea of what it meant at this point, I knew that the standard Buddhist definition of the mental, as opposed to the physical, is characterized by subjectivity. Material objects characteristically have a spatial dimension and can be seen to obstruct other material objects. Mental phenomena, by contrast, must be viewed in terms of their temporal sequence and their experiential nature.

I spent a great deal of time studying the distinctions between sensory experience and mental experience. A defining mark of sensory experience is its contingence upon a specific sense organ—the eye, the ear, and so forth. There is a clear recognition that each sense perception is distinct from the others and has an exclusive domain, so that the eye cannot access sound or the ear taste and so on. As was noted by the early Buddhist thinkers, including Vasubandhu and Dharmakirti, there are significant differences in the spatiotemporal processes involved in apprehending the objects of the various sensory spheres. Visual perception of an object can take place from a great distance, hearing a sound from a lesser distance, while the experience of a particular smell occurs within a still shorter range. In contrast, the remaining two senses—giving rise to gustatory and tactile experience—need direct contact between the senses and their respective objects. In scientific language, I would guess, these differences will be explained in terms of the ways physical entities such as photons and sound waves emitted from objects stimulate the sense organs.

The defining characteristic of mental experience is the lack of a physical sense organ. By mental experience, which is effectively a sixth faculty in addition to the five senses, Buddhist theory of mind does not mean anything cryptic or mysterious. Rather, if one is looking at a beautiful flower, the immediate perception of the flower, with all its richness, color, and shape, belongs to visual experience. If one continues to look, there arise repeated successions of the same visual perception. However, the moment thought occurs while one is looking at the flower—for instance, in focusing on a particular aspect or quality, such as the depth of color or the turn of a petal—one has engaged mental consciousness. Mental consciousness includes the entire gamut of what we call thought processes—including memory, recognition, discrimination, intention, will, conceptual and abstract thinking, and dreams.

Sensory experience is immediate and all-enveloping. We smell the rose, we see the color, we feel the prick of the thorn without any conscious thought entering into the experience. Thought, by contrast, operates selectively, sometimes even seemingly arbitrarily, by homing in on a specific aspect or characteristic of a given phenomenon. While studying the rose, you may find unbidden thoughts entering your mind: the smell is vaguely citrus and refreshing, the color a soothing pale pink, the thorn sharp and to be avoided. In addition, conceptual cognition relates to objects through a medium such as language or concepts. When we see a beautifully colored flower, like the red rhododendrons that cover the hills around Dharamsala in spring, the experience is a rich but undifferentiated one; but when thoughts arise about the characteristics of the flower, such as "it is fragrant" or "its petals are large," then the experience is much narrower but more focused.

An excellent analogy, often given to young students, is holding a cup of tea. Sensory experience is like touching the cup with one's bare hands, while thought is like touching the cup with one's hands covered by a cloth. The qualitative difference between these two experiences is stark. The cloth is a metaphor for the concepts and the language which come between observer and object when thought operates.

There is an extensive philosophical analysis of the role of language in relation to thought in Buddhist epistemology, which developed many of its standpoints within the wider context of philosophical engagement with various non-Buddhist schools of thought. Two of the most influential Buddhist figures were the Indian logicians Dignaga and Dharmakirti in the fifth and seventh centuries. During my training in logic and epistemology, I had to memorize crucial sections of Dharmakirti's famous work *The Exposition of Valid Cognition (Pramanavartika)*, a philosophical treatise composed in verse and noted for its dense, literary style. I gather that Western philosophy has been much concerned with the relationship between language and thought, and with the fundamental question of whether thought is entirely contingent on language. Buddhist thinkers, while acknowledging the intimate connection between the two in human beings, accept in principle the possibility of nonlinguistic thought—for instance, animals are believed to have thoughts mediated by concepts (however rudimentary) but not by language in the sense that we understand it.

I was intrigued when I came to discover that in modern Western psychology there is no developed notion of a non-sensory mental faculty. I gather that for many people the expression "sixth sense" connotes some kind of paranormal psychic ability. But for Buddhists it refers to the mental realm, including thoughts, emotions, intentions, and conceptions. There are in Western thought notions like the soul among theists or ego for psychoanalysts, which fill some of the gap, but what seems to be missing is the recognition of a specific faculty that apprehends mental phenomena. Such phenomena include a wide array of cognitive experiences, such as memory and recollection, which are, from the Buddhist point of view, qualitatively different from sensory experience.

Given that the neurobiological model of perception and cognition is an account of these phenomena in terms of the chemical and biological processes of the brain, I can see why from the scientific point of view no qualitative distinctions need be drawn between sensory and conceptual processes. It turns out that the part of the brain most associated with visual perceptions is also that part most active in imaginative visualization. So far as the brain is concerned, it seems as if it makes no difference whether one is seeing something with one's physical eyes or with the "mind's eye." From the Buddhist point of view, the problem is that this neurobiological account leaves out the most significant ingredient of these mental events—subjective experience.

The classical Buddhist epistemological model gives no prominence to the brain in cognitive activity, such as perception. Given Buddhist philosophy's emphasis on empiricism and given that ancient Indian medical science had detailed knowledge of the human anatomy, it is surprising that there was no clear recognition of the role of the brain as the core organizing structure within the body, especially in relation to perception and cognition. Vajrayana Buddhism, however, speaks of the conduit located at the crown of the head as the primary seat of the energy that regulates subjective experience.

In the study of perception and cognition, I can envision a fruitful collaboration between Buddhism and modern neuroscience. Buddhism has much to learn about the brain mechanisms related to mental events—neurological and chemical processes, the formation of synaptic connections, the correlation between specific cognitive states and specific areas of the brain. In addition, there is much value in the medical and biopharmacological knowledge

currently being generated about how brains function when parts
have been damaged and how certain substances induce particular
states.

At one of the Mind and Life conferences, Francisco Varela
showed me a series of MRI images, horizontal cross sections of a
single brain with parts lit up in different colors to indicate relative
neural and chemical activity associated with various sensory experiences. These images were the results of experiments in which the
subject was exposed to different sensory stimuli (such as music or
visual objects) and then recorded in different responses (such as
with the eyes open or closed). It was very convincing to see the
close correlation between the measurable, visible changes in the
brain and the occurrence of specific sensory perceptions. It is this
level of technical precision and the possibilities that arise from the
use of such instruments that mark the marvelous potential of scientific work. When rigorous third-person investigation is combined with rigorous first-person investigation, we can hope to have
a more comprehensive method of studying consciousness.

According to Buddhist epistemology, there is a built-in limitation in the human mind's capacity for ascertaining its objects. This
limitation is temporal insofar as an ordinary mind, untrained in the
deliberate application of meditative attention, can ascertain only
an event that has lasted over a certain length of time — traditionally
the span of a finger snap or the blink of an eye. Events shorter than
this may be perceived but cannot be wholly registered in that they
are not subject to conscious recollection. A further characteristic of
human perception is our need to hang on to things and events only
in terms of their composite nature. For example, if I look at a vase,
I see a bulbous shape with a flat base and decorations. I do not see
the individual atoms and molecules or the space between them,
which together make up the composite phenomenon I observe. So

when a perception occurs, it is not a case of simple mirroring in the mind of what is outside but rather a complex process of organization that takes place to make sense of what are technically infinite amounts of information.

This process of creative construction operates also on the temporal level. In perceiving an event even for the duration of a finger snap, which in fact is composed of myriad minute temporal sequences, we conflate all these "moments" into one continuum. A good analogy for this, cited by Dharmakirti and taught to students in the Tibetan monastic colleges, is that when at night you turn a flaming torch around in a circle, an observer sees a wheel of fire. If you closely observe the "wheel," you see that it is made up of a series of illuminated moments. Thinking back to my childhood, when I was absorbed in the mechanics of film projectors, I realize that the moving image of a film on a screen is in fact made up of a series of stills. Yet we perceive the movie as fluid motion.

The questions of how perceptions arise and, in particular, what the relationship is between a perceptual event and its objects have been major areas of interest for Indian and Tibetan philosophers. There has been a long-standing debate within Buddhist epistemological thinking on how the perception of a given object arises. Three principal standpoints have emerged. One school of thought upholds the view that just as there are a multiplicity of colors in a multicolored object, there are a multiplicity of perceptions in the visual experience of looking at it. A second position maintains that perception is best compared to the splitting of a hard-boiled egg. When an egg is sliced in half, there are two identical pieces; likewise, when the senses come into contact with their respective objects, a single perceptual event splits into an objective half and a subjective half. The third position, traditionally the preferred standpoint of Tibetan thinkers, argues that, regardless of the multiplicity of facets in a chosen object of perception, the actual perceptual experience is a single unitary event.

One important area of discussion in Buddhist epistemology is the analysis of true and false perceptions. For Buddhism, it is knowledge, or correct insight, that brings about freedom from deluded states of mind, so close attention is paid to understanding what constitutes knowledge. The distinction between true understanding and false understanding becomes, therefore, a major concern. There is a vast analysis of all kinds of perceptual experiences and the variety of causes for perceptual errors. If one stands on a boat sailing down the river and perceives the trees on the bank to be moving, the optical illusion lies in the external condition that the boat is moving. If one suffers from jaundice, one may see even a white object like a conch shell as yellow; here the condition for the illusion is internal. If one sees a coiled rope at dusk in an area known for poisonous snakes, one may perceive it as a snake; in this case the condition of illusion is both internal (that is, one's fear of snakes) and external (that is, in the configuration of the rope and the low visibility).

These are all cases where the illusion is conditioned by very immediate circumstances. But there is a whole category of more complex conditions for false cognitions, such as the belief in an autonomous self or the belief that the self or other conditioned phenomena are permanent. During an experience, there's no way to distinguish between accurate and deluded perception. It's only in hindsight that we can make this distinction. It is in effect the subsequent experiences derived from such cognitions that help determine whether they are valid or invalid. It would be interesting to know whether neuroscience will be able to differentiate between accurate and inaccurate perceptions at the level of brain activity.

On several occasions I have asked this question of neuroscientists. So far, to my knowledge, no such experiment has been conducted. On the phenomenological level, we can discern the process by which our mind can go through transitions between several different, in some cases diametrically opposed, states. For example, let us ask whether it was on the moon or on Mars that Neil Armstrong set his foot in 1969. A person could begin by believing strongly that it was Mars. Then, as a result of hearing about the latest Mars probe, he or she might start to waver in this conviction. Once it becomes clear that no manned mission to Mars has yet taken place, he or she might lean toward the correct conclusion that Neil Armstrong landed on the moon. Finally, as a result of speaking to other people and reading accounts of the Apollo mission, the individual may arrive at the correct answer to the original question. In a case like this, we can see that the mind goes from a state of total error through a state of wavering to correct belief and, finally, to true knowledge.

In general, the Tibetan epistemological tradition enumerates a sevenfold typology of mental states: direct perception, inferential cognition, subsequent cognition, correct assumption, inattentive perception, doubt, and distorted cognition. Young monks must learn the definitions of these seven mental states and their complex interrelations; the benefit of studying these states is that by knowing them one can become much more sensitive to the range and complexity of one's subjective experience. Being familiar with these states makes the study of consciousness more manageable.

Much later in my education came the study of Buddhist psychology as it is systematized by the great Indian thinkers Asanga and Vasubandhu. Today many of the works of these authors have been lost in Sanskrit but, thanks to the great efforts of generations of Tibetan translators and their Indian collaborators, they survive in Tibetan. According to some of my Indian friends who are experts in Sanskrit, the Tibetan translations of these Indian classics are so accurate that one can almost imagine what the original Sanskrit texts may have looked like. Asanga's *Compendium of Higher Knowledge* and his younger brother's *Treasury of Higher Knowledge* (the latter is extant in Sanskrit, while only fragments of the first survive in its Sanskrit original) became the primers on Buddhist psychology in Tibet from a very early stage. They are recognized as the root texts of what the Tibetan tradition refers to as the "Higher Abhidharma" school of Asanga and the "Lower Abhidharma" school of Vasubandhu. It is on the basis of these texts that my knowledge of the nature, classification, and functions of mental processes is principally grounded.

Neither Sanskrit nor classical Tibetan has a word for "emotion" as the concept is used in modern languages and cultures. This is not to say that the idea of emotion does not exist, nor does it imply that Indians and Tibetans do not experience emotions. Just as Westerners feel joy at good news, sadness at a personal loss, and fear in the face of danger, so do the Indians and the Tibetans. Perhaps the reasons for the lack of such a word have to do with the history of philosophical thinking and psychological analysis in India and Tibet. Buddhist psychology did not differentiate cognitive from emotional states in the way Western thought differentiated the passions from reason. From the Buddhist perspective, the distinctions between afflictive and non-afflictive mental states are more important than the difference between cognition and emotions. Discerning intelligence, closely associated with reason, may be afflictive (for example, in the cunning planning of an act of murder), whereas a passionate state of mind, such as overwhelming compassion, may be a highly virtuous, non-afflictive state. Moreover, the emotions of both joy and sorrow may be afflictive or non-afflictive, destructive or beneficial, depending on the context in which they arise.

In Buddhist psychology, an important distinction is drawn between consciousness and the various modalities through which it manifests, for which the technical term in Buddhism is "mental factors." For instance, when I see a friend from a distance, this constitutes a mental episode which may appear as a single event but is in fact a highly complex process. There are five factors universal to all mental events—feeling (in this case pleasant), recognition, engagement, attention, and contact with the object. In this example, there may be additional factors, such as attachment or excitation, depending upon the observer's state of mind at that instant and the particular object that appears. The mental factors should be seen not as separate entities but rather as different aspects, or processes, of the same mental episode, distinguished in terms of their functions. The emotions belong to the category of mental factors, as opposed to the category of consciousness itself.

Although there are many systems of enumeration, the standard list preferred by the Tibetans, which was formulated by Asanga, contains fifty-one key mental factors. In addition to the five universals (feelings, recognition, engagement, attention, and contact), five factors of object discernment—aspiration, attraction, mindfulness, concentration, and insight—are present when the mind ascertains an object. Further, there are eleven wholesome mental factors, which are present when the mind is in a positive state. These are faith or confidence, a sense of shame, conscience (defined as a consideration of others), non-attachment, non-hate (including loving-kindness), non-delusion (including wisdom), vigor, pliancy, heedfulness, equanimity, and non-harmfulness (including compassion). Within this list, we find several that correspond to positive emotions—notably loving-kindness and compassion. Shame and conscience are inter-esting in that the former is about the capacity for feeling sullied by one's own unwholesome deeds or thoughts, while conscience in this context refers to the quality that causes one to refrain from unwhole-some acts or thoughts out of a consideration for others. Both of these therefore have an emotional element.

When we turn to the afflictive mental processes, the list is fuller, largely because these are what need to be purified by the person as-piring to enlightenment in Buddhism. There are six root mental af-flictions: attachment or craving, anger (which includes hate), pride or conceit, ignorance, afflictive doubt, and afflictive views. Of these, the first three have a strong emotional component. Then there are twenty derivative afflictions: wrath, resentment, spite, envy or jeal-ousy, and cruelty (these are derived from anger); meanness, inflated self-esteem, excitement including surprise, concealment of one's own vices, and mental dullness (these are derived from attachment); lack of confidence, sloth, forgetfulness, and lack of attention (these are derived from ignorance); pretentiousness, deceit, shameless-ness, lack of consideration for others, heedlessness, and distraction (these are derived from the combination of ignorance and attach-ment). Clearly many of the mental factors enumerated here can be identified with emotions. Finally, in the list of fifty-one, there is a group of four mental factors referred to as the "changeables." These are sleep, regret, investigation, and minute analysis. They are called changeables because, depending on the state of mind, they can be wholesome, unwholesome, or neutral.

It is most important to be sensitive to the differing contexts in which Buddhist and Western psychology provide treatment of the emotions. We must not confuse the Buddhist distinction between the wholesome and the unwholesome emotions with Western psychol-ogy's distinction between the positive and the negative emotions. In

Western thought, *positive* and *negative* are defined in terms of how one feels when particular emotions occur. For example, fear is negative because it brings about an unpleasant feeling of disturbance.

The Buddhist differentiation between unwholesome or afflictive and wholesome mental factors is based on the roles these factors play in relation to the acts they give rise to—in other words, one's ethical well-being. For instance, attachment may feel enjoyable but is regarded as afflictive since it involves the kind of blind clinging, based on self-centeredness, which can motivate one to harmful action. Fear is neutral and indeed changeable in that it may spur one to wholesome or unwholesome behavior depending on the circumstances. The role of these emotions as motivating factors in human action is highly complex and has attracted wide-ranging attention in the Buddhist treatises. The original Tibetan term for affliction, *nyönmong*, and its Sanskrit equivalent, *klesha*, connote something that afflicts from within. A key characteristic of these mental states is their effect in creating disturbance and a loss of self-control. When they arise, we tend to lose our freedom to act in accordance with our aspirations and become caught in a distorted mind-set. Given that they are ultimately rooted in a deeply self-centered way of relating to others and to the world at large, when these afflictions arise, our perspectives tend to become narrow.

There is extensive analysis of the nature, permutations, subdivisions, interrelationships, and causal dynamics of the mental factors in both Indian and Tibetan works on Buddhist psychology. Asanga's list, which we are using here, must not be considered exhaustive—for instance, fear and anxiety do not appear, although they arise in other contexts and other lists. Despite the differences among the systems of enumeration, the organization of the lists of mental factors reflects the underlying objective of identifying and clearing negative emotions and cultivating positive states of mind.

I have long wondered how we might relate the Buddhist psychological framework of wholesome and unwholesome mental processes to the understanding of emotions developed by Western science. The tenth Mind and Life conference, in March 2000, gave me an opportunity to think about this area more deeply, since the theme of the conference was destructive emotions, and a number of experts on emotion from the Western scientific community came for a weeklong discussion in Dharamsala. The proceedings were moderated by Daniel Goleman, whom I have known for a long time. It was Dan who first presented to me the numerous scientific studies that suggest a close relation between an individual's overall state of mind and his or her physical health. This conference is where I met Paul Ekman, an anthropologist and psychologist who has spent several decades studying emotions. I felt an immediate affinity with him and sensed that a genuine ethical motivation underlies his work, in that if we understand the nature of our emotions and their universality better, we may be able to develop a greater sense of kinship in humanity. Also, Paul speaks at exactly the right pace for me to follow his presentation in English without difficulty.

I learned a great deal from Paul about the latest scientific understanding of emotion. I understand that modern cognitive science draws distinctions between two principal categories of emotions—basic emotions and what some people refer to as "higher cognitive emotions." By "basic emotions," scientists mean those emotions which are thought to be universal and innate. As in Buddhist lists, the precise enumeration differs by researcher, but Ekman mentions as many as ten, including anger, fear, sadness, disgust, contempt, surprise, enjoyment, embarrassment, guilt, and shame. As in the Buddhist mental factors, each of these is seen as representing a family of feelings. By the "higher cognitive emotions," scientists mean a series of emotions that are also universal but whose expression is subject to considerable cultural variation. Examples include love, pride, and jealousy. Experimenters have observed that while basic emotions appear largely to be processed in the subcortical structures of the brain, the higher cognitive emotions are associated more with the neocortex—the part of the brain that has developed most in human evolution and is most active in complex cognitive activity, such as reasoning. I realize that all this represents the very preliminary results of an ongoing and quickly evolving discipline, which may well undergo a radical paradigm shift before it settles into a consensus.

Buddhism assumes the universality of mental afflictions in all sentient beings. The key afflictions are seen as expressions of attachment, anger, and delusion. In some species, such as human beings, the expressions of these are more complex, while in certain species of animals their manifestations will be more rudimentary and more nakedly aggressive. The simpler they are, the more such processes are considered to be instinctual and less dependent upon conscious thinking. In contrast, the complex expressions of emotion are seen as more susceptible to conditioning, including by language and concepts. So the possibility that the basic emotions, according to the classification of modern science, are associated with parts of the brain that are much older in terms of evolution and shared with the animals offers a potential parallel with Buddhism's understanding.

From an experiential point of view, one difference between the afflictive emotions, such as hate, and wholesome states, such as compassion, is that the afflictions tend to fixate the mind on a concrete target—a person to whom we become attached, a smell or sound we want to push away. The wholesome emotions, by contrast, can be more diffuse, so the focus is not confined to one person or one object. There is therefore in Buddhist psychology a notion that the more wholesome mental states have a higher cognitive component than the negative afflictions. Again, this might prove an interesting area of comparison and research with Western science.

Given that the modern science of emotions is grounded in neurobiology, the evolutionary perspective is bound to remain the overarching conceptual frame. This means that, in addition to explorations of the neurological basis of individual emotions, attempts will be made to understand the emergence of specific emotions in terms of their role in natural selection. I am told there is in fact an entire discipline called "evolutionary psychology." To an extent I can see how evolutionary accounts can be given for the emergence of basic emotions such as attachment, anger, and fear. However, as in the neurobiological project that attempts to tie particular emotions to specific areas of the brain, I cannot envision how the evolutionary approach can do justice to the richness of the emotional world and the subjective quality of experience.

Another very interesting point that emerged from my discussions with Paul Ekman is the distinction between emotions on the one hand and moods and traits on the other. Emotions are seen as instantaneous, whereas moods may last longer—even for a whole day—and traits are longer lasting still, sometimes carried for a lifetime. Joy and sadness, for example, would be emotions, which often arise out of a particular stimulus; while happiness and unhappiness would be moods, whose direct causes might not be so easy to identify. Similarly, fear is an emotion, but anxiety is its corresponding mood, while an individual may possess a high propensity for being anxious, which would be a trait of character. Although Buddhist psychology makes no formal distinction between moods and emotions, it recognizes the differences between mental states, both instantaneous and enduring, and the underlying propensities toward them.

The ideas that particular emotions may arise from a certain natural propensity, that specific emotions may give rise to certain kinds of behavior, and in particular the assumption that positive emotions are more amenable to thought processes are critical for Buddhist contemplative practice. Key practices such as the cultivation of compassion and loving-kindness, or the overcoming of destructive emotions such as anger and hatred, are rooted in and dependent upon the insights of psychology. A crucial aspect of these practices is a minute analysis of the causal dynamics of specific mental processes—their external conditions, preceding and attendant internal mental states, and relationship to other cognitive and emotional events. On a number of occasions I have had discussions with psychologists and psychoanalysts in a wide area of therapeutic disciplines, and I have noticed a parallel interest in the causality of emotions. Insofar as these disciplines of applied psychology are concerned with the alleviation of suffering, I think that they share a fundamental goal with Buddhism.

The primary purpose of Buddhist contemplative practice is to alleviate suffering. Science, as we have seen, has contributed greatly to the lessening of suffering, particularly in the physical realm. This is a wonderful pursuit that I hope we will all continue to benefit from. But as science advances further, there is more at stake. Science's power to affect the environment, indeed to change the course of the human species as a whole, has grown great. As a result, for the first time in history, our very survival demands that we begin to consider ethical responsibility not just in the application of science but in the direction of research and development of new realities and technology as well. It is one thing to use the study of neurobiology, psychology, and even Buddhist theory of mind to try to become happier, to change our minds through deliberate cultivation of positive states of mind. But when we begin manipulating genetic codes, both of ourselves and of the natural world in which we live, how much is too much? This is a question that must be considered by scientists as well as the public at large.

9.

ETHICS AND THE
NEW GENETICS

Many of us who have followed the development of the new genetics are aware of the deep public disquiet that is gathering around the topic. This concern has been raised in relation to everything from cloning to genetic manipulation. There has been a worldwide outcry over the genetic engineering of foodstuffs. It is now possible to create new breeds of plants with far higher yields and far lower susceptibility to disease in order to maximize food production in a world where the increasing population needs to be fed. The benefits are obvious and wonderful. Seedless watermelons, apples that have longer shelf lives, wheat and other grains that are immune to pests when growing in the field—these are no longer science fiction. I have read that scientists are even experimenting to develop farm products, such as tomatoes, injected with genes from different species of spiders.

But by doing these things, we are changing the genetic makeup, and do we really know what the long-term impact will be on the species of plants, on the soil, on the environment? There are obvious commercial benefits, but how do we judge what is really useful? The complex web of interdependence that characterizes the environment makes it seem beyond our capacity to predict.

Genetic changes have happened slowly over hundreds of thousands of years of natural evolution. The evolution of the human brain has occurred over millions of years. By actively manipulating the gene, we are on the cusp of forcing an unnaturally quick rate of change in animals and plants as well as our own species. This is not to say that we should turn our backs on developments in this area—it is simply to point out that we must become aware of the awesome implications of this new area of science.

The most urgent questions that arise have to do more with ethics than with science per se, with correctly applying our knowledge and power in relation to the new possibilities opened by cloning, by unlocking the genetic code and other advances. These issues relate to the possibilities for genetic manipulation not only of human beings and animals but also of plants and the environment of which we are all parts. At heart the issue is the relationship between our knowledge and power on the one hand and our responsibility on the other.

Any new scientific breakthrough that offers commercial prospects attracts tremendous interest and investment from both the public sector and private enterprise. The amount of scientific knowledge and the range of technological possibilities are so enormous that the only limitations on what we do may be the results of insufficient imagination. It is this unprecedented acquisition of knowledge and power that places us in a critical position at this time. The higher the level of knowledge and power, the greater must be our sense of moral responsibility.

If we examine the philosophical basis underlying much of human ethics, a clear recognition of the principle that correlates greater knowledge and power with a greater need for moral responsibility serves as a key foundation. Until recently we could say that this principle had been highly effective. The human capacity for moral reasoning has kept pace with developments in human knowledge and its capacities. But with the new era in biogenetic science, the gap between moral reasoning and our technological capacities has reached a critical point. The rapid increase of human knowledge and the technological possibilities emerging in the new genetic science are such that it is now almost impossible for ethical thinking to keep pace with these changes. Much of what is soon going to be possible is less in the form of new breakthroughs or paradigms in science than in the development of new technological options combined with the financial calculations of business and the political and economic calculations of governments. The issue is no longer whether we should or should not acquire knowledge and explore its technological potentials. Rather, the issue is how to use this new knowledge and power in the most expedient and ethically responsible manner.

The area where the impact of the revolution in genetic science may be felt most immediately at present is medicine. Today, I gather, many in medicine believe that the sequencing of the human genome will usher in a new era, in which it may be possible to move beyond a biochemical model of therapy to a genetically based model. Already the very definitions of many diseases are changing as illnesses are found to be genetically programmed into human beings and animals from their conception. While successful gene therapy for some of these conditions may be some way off, it seems no longer beyond the bounds of possibility. Even now, the issue of gene therapy and the associated question of genetic manipulation, especially at the level of the human embryo, are posing grave challenges to our capacity for ethical thinking.

A profound aspect of the problem, it seems to me, lies in the question of what to do with our new knowledge. Before we knew that specific genes caused senile dementia, cancer, or even aging, we as individuals assumed we wouldn't be afflicted with these problems, but we responded when we were. But now, or at any rate very soon, genetics can tell individuals and families that they have genes which may kill or maim them in childhood, youth, or middle age. This knowledge could radically alter our definitions of health and sickness. For example, someone who is healthy at present but has a particular genetic predisposition may come to be marked as "soon to be sick." What should we do with such knowledge, and how do we handle it in a way that is most compassionate? Who should have access to such knowledge, given its social and personal implications in relation to insurance, employment, and relationships, as well as reproduction? Does the individual who carries such a gene have a responsibility to reveal this fact to his or her potential partner in life? These are just a few of the questions raised by such genetic research.

To complicate an already intricate set of problems, I gather that genetic forecasting of this kind cannot be guaranteed to be accurate. It is sometimes certain that a particular genetic disorder observed in the embryo will give rise to disease in the child or adult, but it is often a question of relative probabilities. Lifestyle, diet, and other environmental factors come into play. So while we may know that a particular embryo carries a gene for a disease, we cannot be certain that the disease will arise.

People's life choices and indeed their very self-identity may be significantly affected by their perception of genetic risk, but those perceptions may not be correct and the risk may not be actualized.

Should we be afforded such probabilistic knowledge? In cases where one member of the family discovers a genetic disorder of this type, should all the other members who may have inherited the same gene be informed? Should this knowledge be made available to a wider community—for instance, to health insurance companies? The carriers of certain genes may be excluded from insurance and hence even from access to health care all because there is a possibility of a particular disease manifesting itself. The issues here are not just medical but ethical and can affect the psychological well-being of the people concerned. When genetic disorders are detected in the embryo (as will increasingly be the case), should parents (or society) make the decision to curtail the life of that embryo? This question is further complicated by the fact that new methods of dealing with genetic disease and new medications are being found as swiftly as the genes carrying individual disease are identified. One can imagine a scenario in which a baby whose disease may manifest in twenty years is aborted and a cure for the disease is found within a decade.

Many people around the world, especially practitioners of the newly emerging discipline of bioethics, are grappling with the specifics of these problems. Given my lack of expertise in these fields, I have nothing concrete to offer in regard to any specific question—especially as the empirical facts are changing so rapidly. What I wish to do, however, is think through some of the key issues which I feel every informed person in the world needs to reflect upon, and to suggest some general principles that can be brought to bear in dealing with these ethical challenges. I believe that at heart the challenge we face is really a question of what choices we make in the face of the growing options that science and technology provide us.

Attendant on the new frontiers of genetically based medicine there is a series of further issues which again raise deep and troubling ethical questions. Here I am speaking primarily of cloning. It has now been several years since the world was introduced to a completely cloned sentient being, Dolly, the famous sheep. Since then there has been a huge amount of coverage of human cloning. We know that the first cloned human embryos have been created. The media frenzy aside, the question of cloning is highly complex. I am told there are two quite different kinds of cloning—therapeutic and reproductive. Within therapeutic cloning, there is the use of cloning technology for the reproduction of cells and the potential creation of semi-sentient beings purely for the purpose of harvesting body parts for transplantation. Reproductive cloning is basically the creation of an identical copy.

In principle, I have no objection to cloning as such—as a technological instrument for medical and therapeutic purposes. As in all these cases, what must govern one's decisions is the question of compassionate motivation. However, regarding the idea of deliberately breeding semi-human beings for spare parts, I feel an immediate, instinctive revulsion. I once saw a BBC documentary which simulated such creatures through computer animation, with some distinctively recognizable human features. I was horrified. Some people might feel this is an irrational emotional reaction that need not be taken seriously. But I believe we must trust our instinctive feelings of revulsion, as these arise out of our basic humanity. Once we allow the exploitation of such hybrid semi-humans, what is to stop us from doing the same with our fellow human beings whom the whims of society may deem deficient in some way? The willingness to step across such natural thresholds is what often leads humanity to the commission of horrific atrocities.

Although reproductive cloning is not horrifying in the same way, in some respects its implications may be more far-reaching.

Once the technology becomes feasible, there could be parents who, desperate to have children and unable to do so, may seek to bear a child through cloning. What would this practice do to the future gene pool? To the diversity that has been essential to evolution?

There could also be individuals who, out of a desire to live beyond biological possibility, may choose to clone themselves in the belief that they will continue to live in the new cloned being. In this case, I find it difficult to see any justifiable motives—from the Buddhist perspective, it may be an identical body, but there will be two different consciousnesses. They will still die.

One of the social and cultural consequences of new genetic technologies is their effect on the continuation of our species, through interference with the reproductive process. Is it right to select the sex of one's child, which I believe is possible now? If it is not, is it right to make such choices for reasons of health (say, in couples where a child is at serious risk of muscular dystrophy or hemophilia)? Is it acceptable to insert genes into human sperm or eggs in the lab? How far can we go in the direction of creating "ideal" or "designer" fetuses—for instance, embryos that have been selected in the lab to provide particular molecules or compounds absent in genetically deficient siblings in order that the children born from such embroys may donate bone marrow or kidneys to cure siblings? How far can we go with the artificial selection of fetuses with desirable traits that are held to improve intelligence or physical strength or specific color of eyes for instance?

When such technologies are used for medical reasons—as in the curing of a particular genetic deficiency—one can deeply sympathize. The selection of particular traits, however, especially when done for primarily aesthetic purposes, may not be for the benefit of the child. Even when the parents think they are selecting traits that will positively affect their child, we need to consider whether this is being done out of positive intention or on the basis of a particular society's prejudices at a particular time. We have to bear in mind the long-term impact of this kind of manipulation on the species as a whole, given that its effects will be passed on to following generations. We need also to consider the effects of limiting the diversity of humanity and the tolerance that goes with it, which is one of the marvels of life.

Particularly worrying is the manipulation of genes for the creation of children with enhanced characteristics, whether cognitive or physical. Whatever inequalities there may be between individuals in their circumstances—such as wealth, class, health, and so on—we are all born with a basic equality of our human nature, with certain potentialities; certain cognitive, emotional, and physical abilities; and the fundamental disposition—indeed the right—to seek happiness and overcome suffering. Given that genetic technology is bound to remain costly, at least for the foreseeable future, once it is allowed, for a long period it will be available only to a small segment of human society, namely the rich. Thus society will find itself translating an inequality of circumstance (that is, relative wealth) into an inequality of nature through enhanced intelligence, strength, and other faculties acquired through birth.

The ramifications of this differentiation are far-reaching—on social, political, and ethical levels. At the social level, it will reinforce—even perpetuate—our disparities, and it will make their reversal much more difficult. In political matters, it will breed a ruling elite, whose claims to power will be invocations of an intrinsic natural superiority. On the ethical level, these kinds of pseudo-nature-based differences can severely undermine our basic moral sensibilities insofar as these sensibilities are based on a mutual recognition of shared humanity. We cannot imagine how such practices could affect our very concept of what it is to be human.

When I think about the various new ways of manipulating human genetics, I can't help but feel that there is something profoundly lacking in our appreciation of what it is to cherish humanity. In my native Tibet, the value of a person rests not on physical appearance, not on intellectual or athletic achievement, but on the basic, innate capacity for compassion in all human beings. Even modern medical science has demonstrated how crucial affection is for human beings, especially during the first few weeks of life. The simple power of touch is critical for the basic development of the brain. In regard to his or her value as a human being, it is entirely irrelevant whether an individual has some kind of disability—for instance, Down syndrome—or a genetic disposition to develop a particular disease, such as sickle-cell anemia, Huntington's chorea, or Alzheimer's. All human beings have an equal value and an equal potential for goodness. To ground our appreciation of the value of a human being on genetic makeup is bound to impoverish humanity, because there is so much more to human beings than their genomes.

For me, one of the most striking and heartening effects of our knowledge of the genome is the astounding truth that the differences in the genomes of the different ethnic groups around the world are so negligible as to be insignificant. I have always argued that the differences of color, language, religion, ethnicity, and so forth among human beings have no substance in the face of our basic sameness. The sequencing of the human genome has, for me, demonstrated this in an extremely powerful way. It has also helped reinforce my sense of our basic kinship with animals, who share very large percentages of our genome. So it is conceivable if we humans utilize our newly found genetic knowledge skillfully, it could help foster a greater sense of affinity and unity not only with our fellow human beings but with life as a whole. Such a perspective could also underpin a much more healthy environmental consciousness.

In the case of food, if the argument is valid that we need some kind of genetic modification to help feed the world's growing population, then I believe that we cannot simply dismiss this branch of genetic technology. However, if, as suggested by its critics, this argument is merely a front for motives that are primarily commercial—such as producing food that will simply have a longer lasting shelf life, that can be more easily exported from one side of the world to the other, that is more attractive in appearance and more convenient in consumption, or creating grains and cereals engineered not to produce their own seeds so that farmers are forced to depend entirely upon the biotech companies for seeds—then clearly such practices must be seriously questioned.

Many people are becoming increasingly worried by the long-term consequences of producing and consuming genetically modified produce. The gulf between the scientific community and the general public may be caused in part by the lack of transparency in the companies developing these products. The onus should be on the biotech industry both to demonstrate that there are no long-term negative consequences for consumers of these new products and to adopt complete transparency on all the possible implications such plants may have for the natural environment. Clearly the argument that if there is no conclusive evidence that a particular product is harmful then there is nothing wrong with it cannot be accepted.

The point is that genetically modified food is not just another product, like a car or a portable computer. Whether we like it or not, we do not know the long-term consequences of introducing genetically modified organisms into the wider environment. In medicine, for instance, the drug thalidomide was found to be excellent for the treatment of morning sickness in pregnant women, but its long-term consequences for the health of the unborn child were not foreseen and proved catastrophic.

Given the tremendous pace of development in modern genetics, it is urgent now to refine our capacity for moral reasoning so that we are equipped to address the ethical challenges of this new situation. We cannot wait for a series of responses to emerge in an organic way. We need to confront the reality of our potential future and tackle the problems directly.

I feel the time is ripe to engage with the ethical side of the genetic revolution in a manner that transcends the doctrinal standpoints of individual religions. We must rise to the ethical challenge as members of one human family, not as a Buddhist, a Jew, a Christian, a Hindu, a Muslim. Nor is it adequate to address these ethical challenges from the perspective of purely secular, liberal political ideals, such as individual freedom, choice, and fairness. We need to examine the questions from the perspective of a global ethics that is grounded in the recognition of fundamental human values that transcend religion and science.

It is not adequate to adopt the position that our responsibility as a society is simply to further scientific knowledge and enhance our technological power. Nor is it sufficient to argue that what we do with this knowledge and power should be left to the choices of individuals. If this argument means that society at large should not interfere with the course of research and the creation of new technologies based on such research, it would effectively rule out any significant role for humanitarian or ethical considerations in the regulation of scientific development. It is essential, indeed it is a responsibility, for us to be much more critically self-aware about what we are developing and why. The basic principle is that the earlier one intervenes in the causal process, the more effective is one's prevention of undesirable consequences.

In order to respond to the challenges in the present and in the future, we need a much higher level of collective effort than has been seen yet. One partial solution is to ensure that a larger segment of the general public has a working grasp of scientific thinking and an understanding of key scientific discoveries, especially those which have direct social and ethical implications. Education needs to provide not only training in the empirical facts of science but also an examination of the relationship between science and society at large, including the ethical questions raised by new technological possibilities. This educational imperative must be directed at scientists as well as laypeople, so that scientists retain a wider understanding of the social, cultural, and ethical ramifications of the work they are doing.

Given that the stakes for the world are so high, the decisions about the course of research, what to do with our knowledge, and what technological possibilities should be developed cannot be left in the hands of scientists, business interests, or government officials. Clearly, as a society we need to draw some lines. But these deliberations cannot come solely from small committees, no matter how august or expert they may be. We need a much higher level of public involvement, especially in the form of debate and discussion, whether through the media, public consultation, or the action of grassroots pressure groups.

Today's challenges are so great—and the dangers of the misuse of technology so global, entailing a potential catastrophe for all humankind—that I feel we need a moral compass we can use collectively without getting bogged down in doctrinal differences. One key factor that we need is a holistic and integrated outlook at the

level of human society that recognizes the fundamentally interconnected nature of all living beings and their environment. Such a
moral compass must entail preserving our human sensitivity and
will depend on us constantly bearing in mind our fundamental human values. We must be willing to be revolted when science—or
for that matter any human activity—crosses the line of human de-
cency, and we must fight to retain the sensitivity that is otherwise
so easily eroded.

How can we find this moral compass? We must begin by put-
ting faith in the basic goodness of human nature, and we need to
anchor this faith in some fundamental and universal ethical principles. These include a recognition of the preciousness of life, an
understanding of the need for balance in nature and the employment of this need as a gauge for the direction of our thought and
action, and—above all—the need to ensure that we hold compassion as the key motivation for all our endeavors and that it is combined with a clear awareness of the wider perspective, including
long-term consequences. Many will agree with me that these ethical values transcend the dichotomy of religious believers and non-
believers, and are crucial for the welfare of all humankind. Because
of the profoundly interconnected reality of today's world, we need
to relate to the challenges we face as a single human family rather
than as members of specific nationalities, ethnicities, or religions.
In other words, a necessary principle is a spirit of oneness of the
entire human species. Some might object that this is unrealistic.
But what other option do we have?

I firmly believe it is possible. The fact that, despite our living
for more than half a century in the nuclear age, we have not yet annihilated ourselves is what gives me great hope. It is no more coincidence that, if we reflect deeply, we find these ethical principles at
the heart of all major spiritual traditions.

In developing an ethical strategy with respect to the new genetics, it is vitally important to frame our reflection within the widest possible context. We must first of all remember how new this field is and how new are the possibilities it offers, and to contemplate how little we understand what we know. We have now sequenced the whole of the human genome, but it may take decades for us fully to understand the functions of all the individual genes and their interrelationships, let alone the effects of their interaction with the environment. Too much of our current focus is on the feasibility of a particular technique, its immediate or short-term results and side effects, and what effect it may have on individual liberty. These are all valid concerns, but they are not sufficient. Their purview is too narrow, given that the very conception of human nature is at stake. Because of the far-reaching scope of these innovations, we need to examine all areas of human existence where genetic technology may have lasting implications. The fate of the human species, perhaps of all life on this planet, is in our hands. In the face of the great unknown, would it not be better to err on the side of caution than to transform the course of human evolution in an irreversibly damaging direction?

In a nutshell, our ethical response must involve the following key factors. First, we have to check our motivation and ensure that its foundation is compassion. Second, we must relate to any problem before us while taking into account the widest possible perspective, which includes not only situating the issue within the picture of wider human enterprise but also taking due regard of both short-term and long-term consequences. Third, when we apply our reason in addressing a problem, we have to be vigilant in ensuring that we remain honest, self-aware, and unbiased; the danger otherwise is that we may fall victim to self-delusion. Fourth, in the face of any real ethical challenge, we must respond in a spirit of humility, recognizing not only the limits of our knowledge (both collective and personal) but also our vulnerability to being misguided in the context of such a rapidly changing reality. Finally, we must all—scientists and society at large—strive to ensure that whatever new course of action we take, we keep in mind the primary goal of the well-being of humanity as a whole and the planet we inhabit.

The earth is our only home. As far as current scientific knowledge is concerned, this may be the only planet that can support life. One of the most powerful visions I have experienced was the first photograph of the earth from outer space. The image of a blue planet floating in deep space, glowing like the full moon on a clear night, brought home powerfully to me the recognition that we are indeed all members of a single family sharing one little house. I was flooded with the feeling of how ridiculous are the various disagreements and squabbles within the human family. I saw how futile it is to cling so tenaciously to the differences that divide us. From this perspective one feels the fragility, the vulnerability of our planet and its limited occupation of a small orbit sandwiched between Venus and Mars in the vast infinity of space. If we do not look after this home, what else are we charged to do on this earth?

CONCLUSION

·

SCIENCE, SPIRITUALITY, AND HUMANITY

Looking back over my seventy years of life, I see that my personal encounter with science began in an almost entirely prescientific world where the technological seemed miraculous. I suppose my fascination for science still rests in an innocent amazement at the wonders of what it can achieve. From these beginnings my journey into science has led me into issues of great complexity, such as science's impact on our understanding of the world, its power to transform human lives and the very earth we live on, and the awesome moral dilemmas which its new findings have posed. Yet one cannot and should not forget the wonder and the beauty of what has been made possible.

The insights of science have enriched many aspects of my own Buddhist worldview. Einstein's theory of relativity, with its vivid thought experiments, has given an empirically tested texture to my grasp of Nagarjuna's theory of the relativity of time. The extraordinarily detailed picture of the behavior of subatomic particles at the minutest levels imaginable brings home the Buddha's teaching on the dynamically transient nature of all things. The discovery of the genome all of us share throws into sharp relief the Buddhist view of the fundamental equality of all human beings.

What is the place of science in the totality of human endeavor? It has investigated everything from the smallest amoeba to the complex neurobiological system of human beings, from the creation of the universe and the emergence of life on earth to the very nature of matter and energy. Science has been spectacular in exploring reality. It has not only revolutionized our knowledge but opened new avenues of knowing. It has begun to make inroads into the complex question of consciousness—the key characteristic that makes us sentient. The question is whether science can provide a comprehensive understanding of the entire spectrum of reality and human existence.

From the Buddhist perspective, a full human understanding must not only offer a coherent account of reality, our means of apprehending it, and the place of consciousness but also include a clear awareness of how we should act. In the current paradigm of science, only knowledge derived through a strictly empirical method underpinned by observation, inference, and experimental verification can be considered valid. This method involves the use of quantification and measurement, repeatability, and confirmation by others. Many aspects of reality as well as some key elements of human existence, such as the ability to distinguish between good and evil, spirituality, artistic creativity—some of the things we most value about human beings—inevitably fall outside the scope of the method. Scientific knowledge, as it stands today, is not complete. Recognizing this fact, and clearly recognizing the limits of scientific knowledge, I believe, is essential. Only by such recognition can we genuinely appreciate the need to integrate science within the totality of human knowledge. Otherwise our conception of the world, including our own existence, will be limited to the facts adduced by science, leading to a deeply reductionist, materialistic, even nihilistic worldview.

My difficulty is not with reductionism as such. Indeed, many of our great advances have been made by applying the reductionist approach that characterizes so much scientific experimentation and analysis. The problem arises when reductionism, which is essentially a method, is turned into a metaphysical standpoint. Understandably this reflects a common tendency to conflate the means with the end, especially when a specific method is highly effective. In a powerful image, a Buddhist text reminds us that when someone points his finger at the moon, we should direct our gaze not at the tip of the finger but at the moon to which it is pointing.

Throughout this book, I hope I have made the case that one can take science seriously and accept the validity of its empirical findings without subscribing to scientific materialism. I have argued for the need for and possibility of a worldview grounded in science, yet one that does not deny the richness of human nature and the validity of modes of knowing other than the scientific. I say this because I believe strongly that there is an intimate connection between one's conceptual understanding of the world, one's vision of human existence and its potential, and the ethical values that guide one's behavior. How we view ourselves and the world around us cannot help but affect our attitudes and our relations with our fellow beings and the world we live in. This is in essence a question of ethics.

Scientists have a special responsibility, a moral responsibility, in ensuring that science serves the interests of humanity in the best possible way. What they do in their specific disciplines has the power to affect the lives of all of us. For whatever historical reasons, scientists have come to enjoy a much higher level of public trust than other professionals. It is true, however, that this trust is no longer an absolute faith. There have been too many tragedies related either directly or indirectly to science and technology for the trust in science to remain unconditional. In my own lifetime, we need only think of Hiroshima, Chernobyl, Three Mile Island, or Bhopal in terms of nuclear or chemical disasters, and of the degradation of the environment—including the depletion of the ozone layer—among ecological crises.

My plea is that we bring our spirituality, the full richness and simple wholesomeness of our basic human values, to bear upon the course of science and the direction of technology in human society. In essence, science and spirituality, though differing in their approaches, share the same end, which is the betterment of humanity. At its best, science is motivated by a quest for understanding to help lead us to greater flourishing and happiness. In Buddhist language, this kind of science can be described as wisdom grounded in and tempered by compassion. Similarly, spirituality is a human journey into our internal resources, with the aim of understanding who we are in the deepest sense and of discovering how to live according to the highest possible idea. This too is the union of wisdom and compassion.

Since the emergence of modern science, humanity has lived through an engagement between spirituality and science as two important sources of knowledge and well-being. Sometimes the relationship has been a close one—a kind of friendship—while at other times it has been frosty, with many finding the two to be incompatible. Today, in the first decade of the twenty-first century, science and spirituality have the potential to be closer than ever, and to embark upon a collaborative endeavor that has far-reaching potential to help humanity meet the challenges before us. We are all in this together. May each of us, as a member of the human family, respond to the moral obligation to make this collaboration possible. This is my heartfelt plea.